Calculus 3 Review in Bite-Size Pieces

By Kathryn Paulk
Copyright © 2023

Updated: 09/01/2025

Table of Contents

Introduction

This book is a review for students who are currently taking or have already taken a third course in calculus. Calculus 3 topics are presented in short bite-size pieces. Detailed examples are included.

This book has been formatted so that it is easy to read on both paperback and also on electronic devices with the Kindle app (laptop, iPad, Kindle, E-reader, and iPhone).

Calculus 3 Review

A review of Calculus 3 topics is included in the following sections. Key equations are listed, followed by detailed examples to reinforce the topic.

Vectors in 3D

3D Coordinate System

3D Coordinate System

1D	Points in one dimension are represented on a single line, usually called the x-axis. To locate a single point on the x-axis, only one value is needed.
2D	Points in two dimensions are represented by two dimensions (directions), usually called the x and y axes. Together, the x and y axis form a flat plane. To locate a single point on a flat plane, two values are needed. One value tells the location along the x-axis and the second value tells the location along the y-axis.
3D	Points in three dimensions are represented by three dimensions (directions), usually called the x, y, and z axes. Together, the x, y, and z axes form a three-dimensional cube. To locate a single point within the cube, three values are needed. Each value tells the location along one of the edges of the cube.

1, 2, and 3 Dimensional Coordinate Systems

$\mathbb{R}^1$	1D coordinate system. Points: $x = a$
$\mathbb{R}^2$	2D rectangular coord. system. Points: $(x, y) = (a, b)$
$\mathbb{R}^3$	3D rectangular coord. system. Points: $(x, y, z) = (a, b, c)$
	$\mathbb{R}^3\ =\ $ Cartesian Product $\mathbb{R}^3\ =\ \mathbb{R} \times \mathbb{R} \times \mathbb{R}$ $\mathbb{R}^3\ =\ \{\,(x, y, z) \mid x, y, z \in \mathbb{R}\,\}$

Note	$\mathbb{R}\ =\ $ The set of real numbers

Points in 1D -- Examples

Points in 1D are represented on a single line, usually called the x-axis.

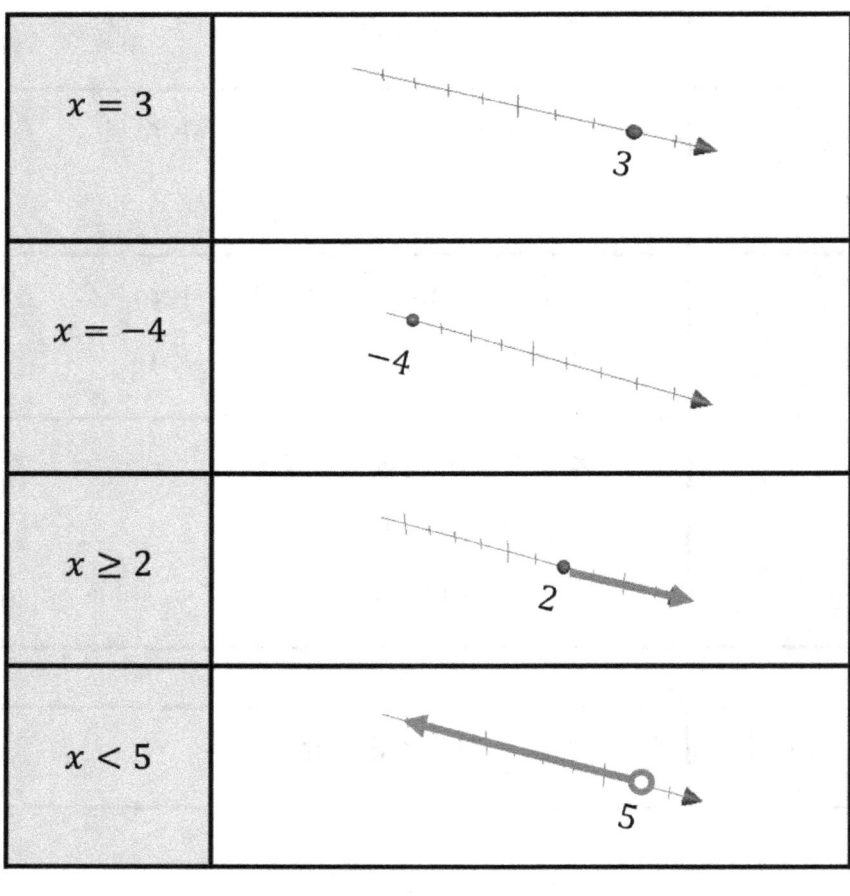

$x = 3$	
$x = -4$	
$x \geq 2$	
$x < 5$	

Points in 2D -- Examples

Points in 2D are represented by 2 dimensions, called the x and y-axis

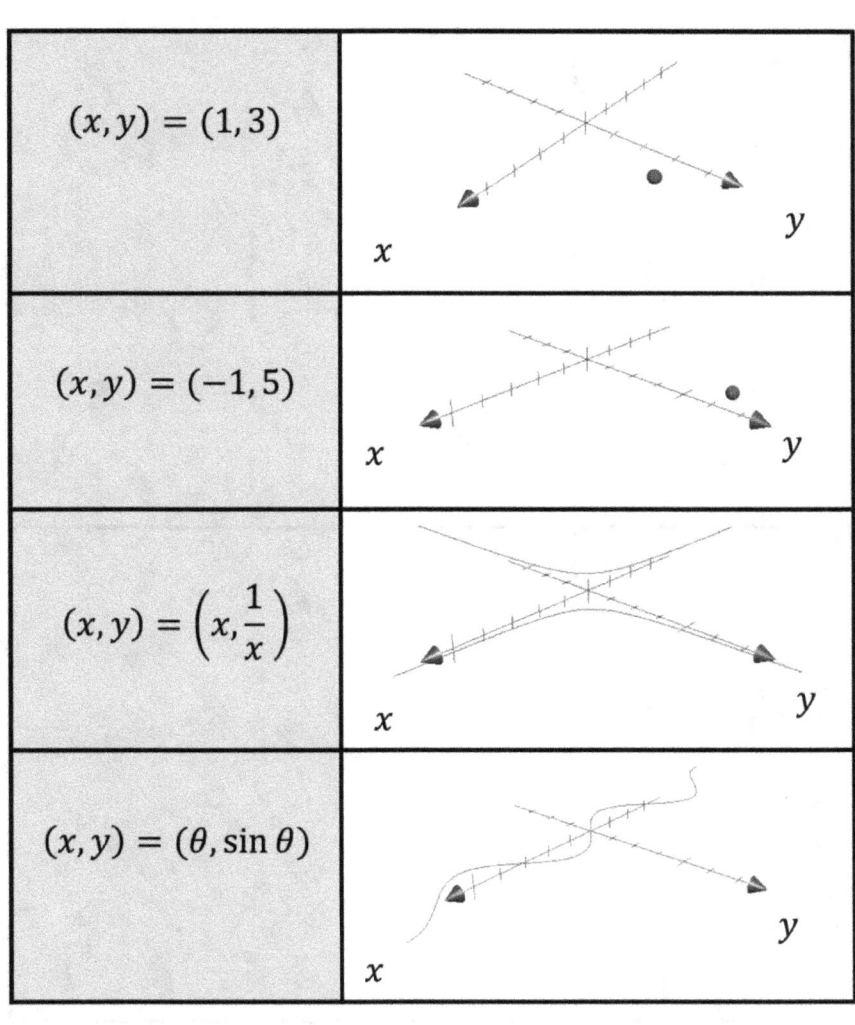

$(x,y) = (1,3)$	
$(x,y) = (-1,5)$	
$(x,y) = \left(x, \dfrac{1}{x}\right)$	
$(x,y) = (\theta, \sin\theta)$	

Points in 3D -- Examples

Points in 3D are represented by 3 dimensions, called the x, y, and z-axis.

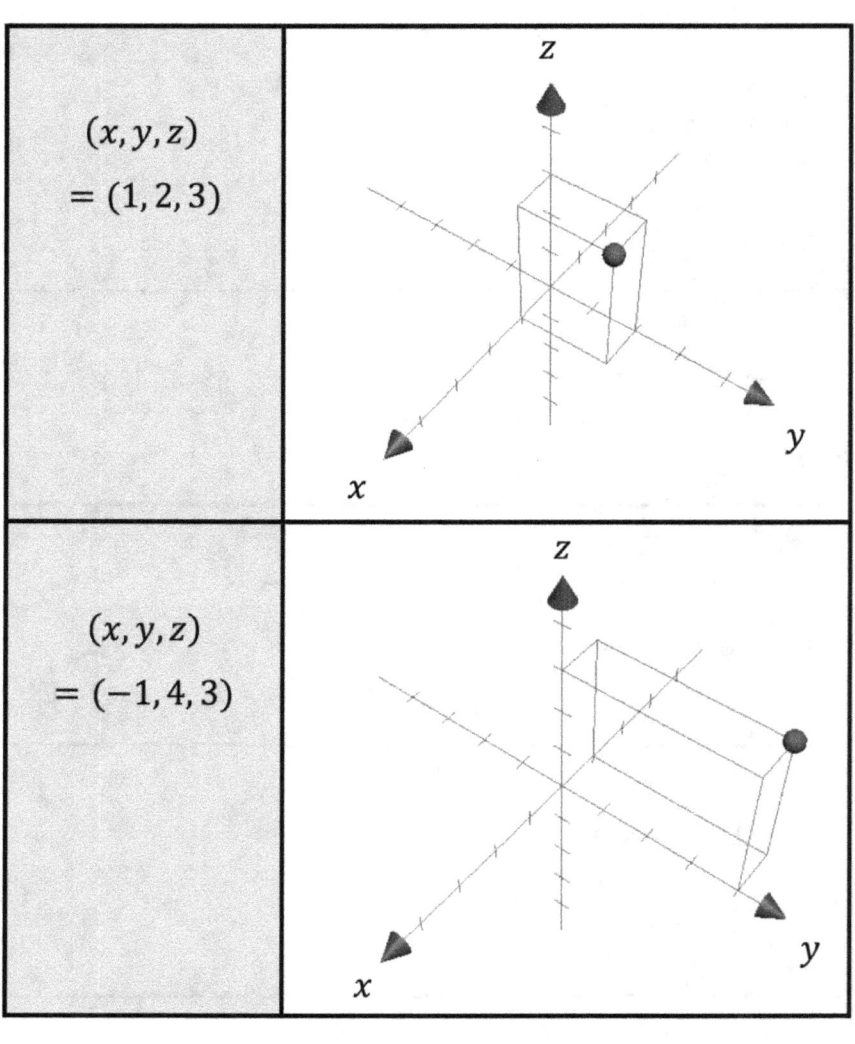

(x, y, z)

$= (1, 2, 3)$

(x, y, z)

$= (-1, 4, 3)$

Surfaces in $\mathbb{R}^3$ -- Examples

In 2D, the graph of an equation involving two variables is a curve in $\mathbb{R}^2$.

In 3D, the graph of an equation involving three variables is a surface in $\mathbb{R}^3$.

$z = 6$ In $\mathbb{R}^3$	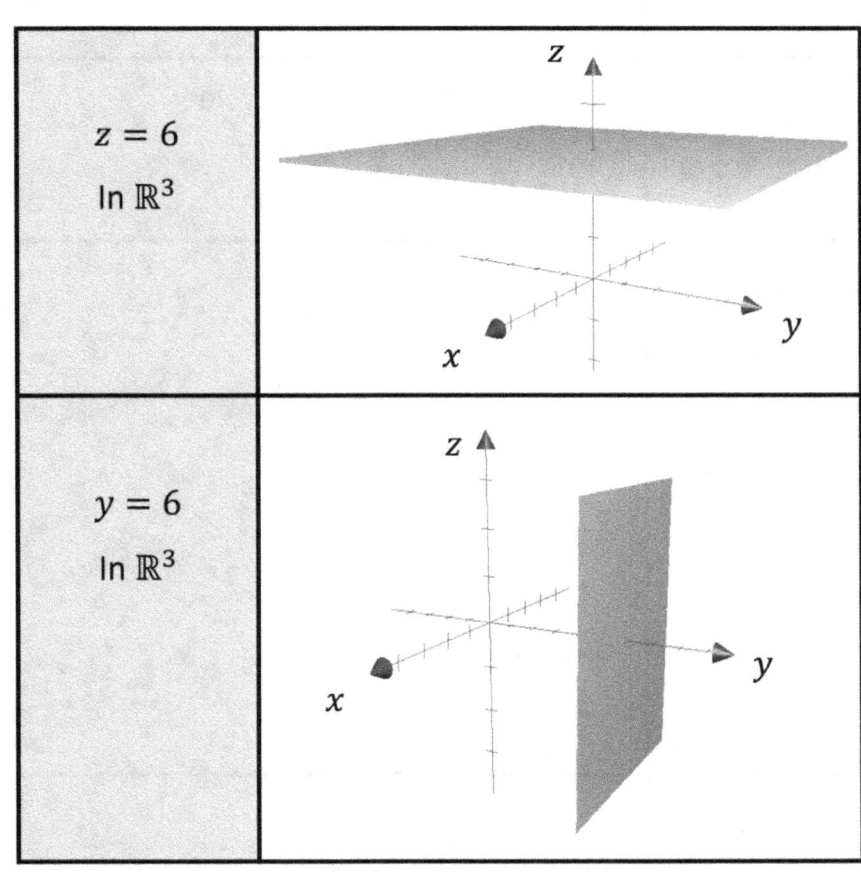
$y = 6$ In $\mathbb{R}^3$	

Curve in $\mathbb{R}^3$ – Circle Example

Sketch the curve in $\mathbb{R}^3$

$$x^2 + y^2 = 1 \quad \text{and} \quad z = 2$$

Note	Restriction on z $z = 2$
Visualize the circle.	A horizontal circle at a height of $z = 2$
Sketch	

Surfaces in $\mathbb{R}^3$	$y = \sin x$

CURVE $y = \sin x$ $-2\pi \leq x \leq 2\pi$ $z = 0$	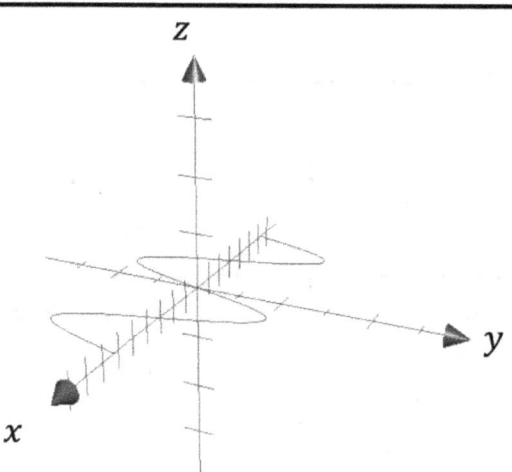
SURFACE $y = \sin x$ $-2\pi \leq x \leq 2\pi$ $0 \leq z \leq 2$	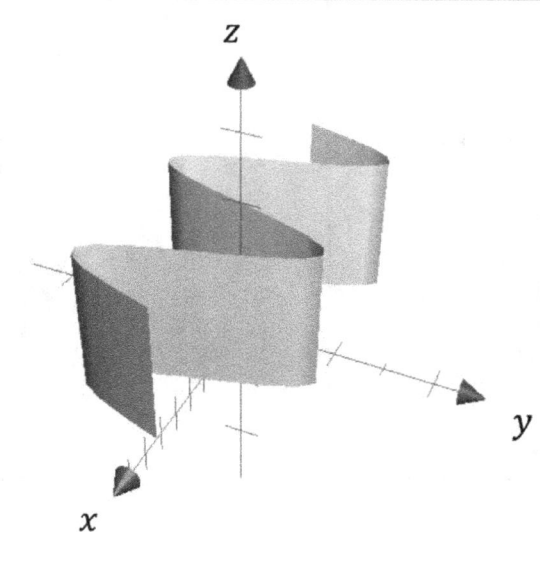

Surfaces in $\mathbb{R}^3$ -- Cylinder	
Sketch the surface in $\mathbb{R}^3$ $x^2 + y^2 = 1$	

Note	No restrictions on z Let $z = k$ (any value)
Visualize a cylinder	All circles where $x^2 + y^2 = 1$ At any height, $z = k$
Sketch	

Surfaces in $\mathbb{R}^3$	$x = y$

Sketch the surface in $\mathbb{R}^3$

$$x = y$$

Note	No restrictions on z Let $z = k$ (any value)
Visualize a plane	All lines where $x = y$ At any height, $z = k$
Sketch Sketch shows $0 \le z \le 3$ and $0 \le x \le 2$	

Distance Formula in $\mathbb{R}^3$

The distance between 2 points

$P_1\,(x_1, y_1, z_1)$ and $P_2\,(x_2, y_2, z_2)$ is:

$$|P_1\,P_2| = \sqrt{(\Delta x)^2 + (\Delta y)^2 + (\Delta z)^2}$$

$$|P_1\,P_2| = \sqrt{(x_2 - x_1)^2 + (y_2 - y_1)^2 + (z_2 - z_1)^2}$$

Distance in $\mathbb{R}^3$ -- Example

Find the distance between 2 points.

$P\,(1, 2, 3)$ and $Q\,(4, 6, 8)$

$$|PQ| = \sqrt{(\Delta x)^2 + (\Delta y)^2 + (\Delta z)^2}$$

$$= \sqrt{(4 - 1)^2 + (6 - 2)^2 + (8 - 3)^2}$$

$$= \sqrt{9 + 16 + 25}$$

$$= \sqrt{50} \quad = \quad 5\sqrt{2}$$

Sphere Equation in $\mathbb{R}^3$

Find the equation of sphere with

r = radius   and   center at $C\,(h, k, l)$

A sphere is the set of all points $P(x, y, z)$

Where distance from the center is r

So $|PC| = r$

$$|PC| = \sqrt{(\Delta x)^2 + (\Delta y)^2 + (\Delta z)^2}$$

$$r = \sqrt{(x - h)^2 + (y - k)^2 + (z - l)^2}$$

$$r^2 = (x - h)^2 + (y - k)^2 + (z - l)^2$$

Note: If the center is at the origin, then

$$r^2 = x^2 + y^2 + z^2$$

Sphere Equation in $\mathbb{R}^3$ -- Example
Show that the given equation is a sphere and find it's center and radius. $$x^2 + y^2 + z^2 + 2x - 4y + 6z = 10$$

Start by completing the square for x, y & z.
Recall: To complete the square, do this: • Make sure coefficient of x^2 is 1. • Divide the coefficient of x by 2. • Then square it. • Add that number to both sides of eqn.
$(x^2 + 2x + 1) + (y^2 - 4y + 4) +$ $+ (z^2 + 6z + 9) \ = \ 10 + 1 + 4 + 9$ $(x + 1)^2 + (y - 2)^2 + (z + 3)^2 \ = \ 24$
Center at: $(x, y, z) = (-1,\ 2, -3)$ Radius: $r = \sqrt{24} = 2\sqrt{6}$

Solid Region in $\mathbb{R}^3$ -- Example

What is the region in $\mathbb{R}^3$ represented by the following inequalities?

$$1 \leq x^2 + y^2 + z^2 \leq 9 \quad \text{and} \quad z \leq 0$$

$x^2 + y^2 + z^2 = 9$	Sphere, centered at origin with radius 3.
$x^2 + y^2 + z^2 = 1$	Sphere, centered at origin with radius 1.
$z \leq 0$	The portion of R3 under the xy plane.

Find Equation in $\mathbb{R}^3$ -- Example

Find the equation of the set of all points equidistant
from two points:

$$A(-1, 3, 5) \quad \text{and} \quad B(2, 4, 0)$$

Distance from A = Distance from B

$$\sqrt{(x+1)^2 + (y-3)^2 + (z-5)^2} \quad =$$
$$\sqrt{(x-2)^2 + (y-4)^2 + (z-0)^2}$$

$$(x+1)^2 + (y-3)^2 + (z-5)^2 =$$
$$(x-2)^2 + (y-4)^2 + (z-0)^2$$

$$x^2 + 2x + 1 + y^2 - 6y + 9 + z^2 - 10z + 25$$
$$= \quad x^2 - 4x + 4 + y^2 - 8y + 16 + z^2$$

$$2x + 1 - 6y + 9 - 10z + 25$$
$$= \quad -4x + 4 - 8y + 16$$

$$6x + 2y - 10z \quad = \quad -15$$

A plane perpendicular to AB

Find Distances in $\mathbb{R}^3$ -- Examples

Find the distance from point $(2, -1, 8)$

To the following ...

xy Plane	Closest point on xy plane: $(2, -1, 0)$ $D = \sqrt{(0)^2 + (0)^2 + (8)^2} = 8$
yz Plane	Closest point on yz plane: $(0, -1, 8)$ $D = \sqrt{(2)^2 + (0)^2 + (0)^2} = 2$
x Axis	Closest point on x axis: $(2, 0, 0)$ $D = \sqrt{(0)^2 + (-1)^2 + (8)^2} = \sqrt{65}$
y Axis	Closest point on y axis: $(0, -1, 0)$ $D = \sqrt{(2)^2 + (0)^2 + (8)^2} = \sqrt{68}$
z Axis	Closest point on z axis: $(0, 0, 8)$ $D = \sqrt{(2)^2 + (-1)^2 + (0)^2} = \sqrt{5}$

Describe Equations in $\mathbb{R}^3$ -- Examples

$0 \leq z \leq 5$	All points between two horizontal planes: $z = 0$ & $z = 5$
$x^2 + z^2 \leq 16$	All points within a circular cylinder in the xz-plane, with $r = 4$. Centered on y-axis.
$y^2 = 4$	$y = \pm\sqrt{4} = \pm 2$ Two parallel planes. $y = 2$ and $y = -2$
$1 \leq x^2 + y^2 + z^2 \leq 4$	All points between spheres with radii of 1 and 2. Centers at $(0,0,0)$

<u>Vectors</u>

Vectors -- Definition
Vectors have magnitude and direction. Vectors are represented as listed below.

$\vec{v}$ or $\boldsymbol{v}$	Vector v
$\overrightarrow{AB}$	Vector from point A to point B
$\boldsymbol{i}$	Unit vector in x direction
$\boldsymbol{j}$	Unit vector in y direction
$\boldsymbol{k}$	Unit vector in z direction
$\langle a, b, c \rangle$	Vector with direction: • a units in x direction • b units in y direction • c units in z direction
$\langle a, b, c \rangle$	$= a\,\boldsymbol{i} + b\,\boldsymbol{j} + c\,\boldsymbol{k}$

Vector -- Illustration

Two points and the vector $\overrightarrow{AB}$

$$A = (\,0,0,0\,) \qquad B = (\,2,6,4\,)$$

$$\overrightarrow{AB} \;=\; \langle\,2,6,4\,\rangle$$

$$\overrightarrow{AB} \;=\; 2\boldsymbol{i} + 6\boldsymbol{j} + 4\boldsymbol{k}$$

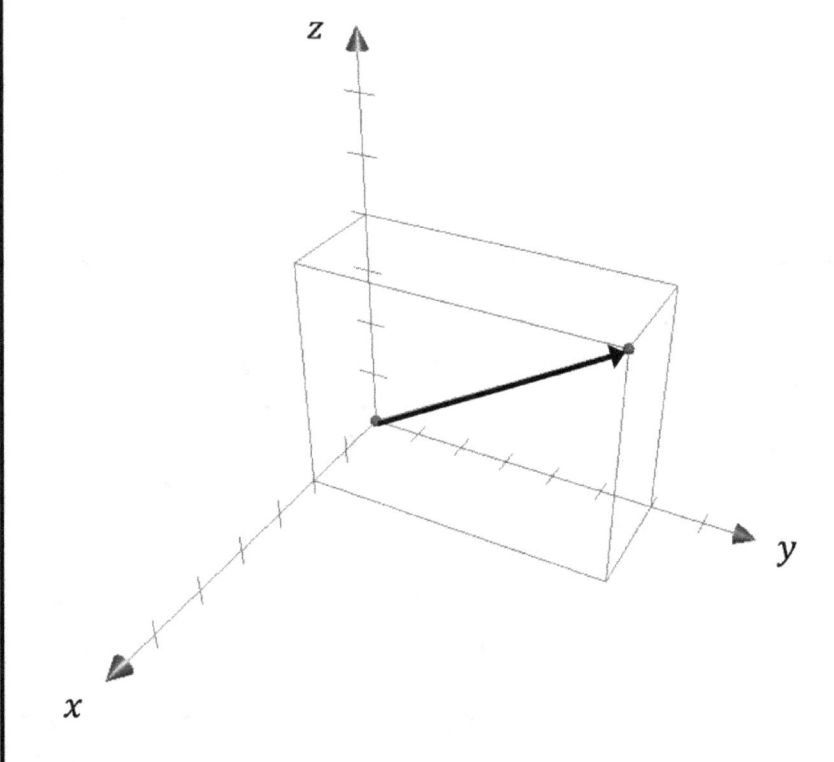

Vectors -- Add & Subtract

To add or subtract two vectors, simply add or subtract the component vectors.

$$a = <a_1, a_2, a_3>$$

$$b = <b_1, b_2, b_3>$$

$$a + b = \langle a_1 + b_1,\ a_2 + b_2,\ a_3 + b_3 \rangle$$

$$a - b = \langle a_1 - b_1,\ a_2 - b_2,\ a_3 - b_3 \rangle$$

Vector Addition -- Example

$$a = <1, 2, 3> \quad \text{and} \quad b = <-2, 5, 10>$$

$$a + b = \langle 1 - 2,\ 2 + 5,\ 3 + 10 \rangle$$

$$a + b = \langle -1,\ 7,\ 13 \rangle$$

Vector Addition -- $\mathbb{R}^2$ Illustration
Two vectors, $\boldsymbol{u}$ and $\boldsymbol{v}$ Added "tip to tail"

The Triangle Law	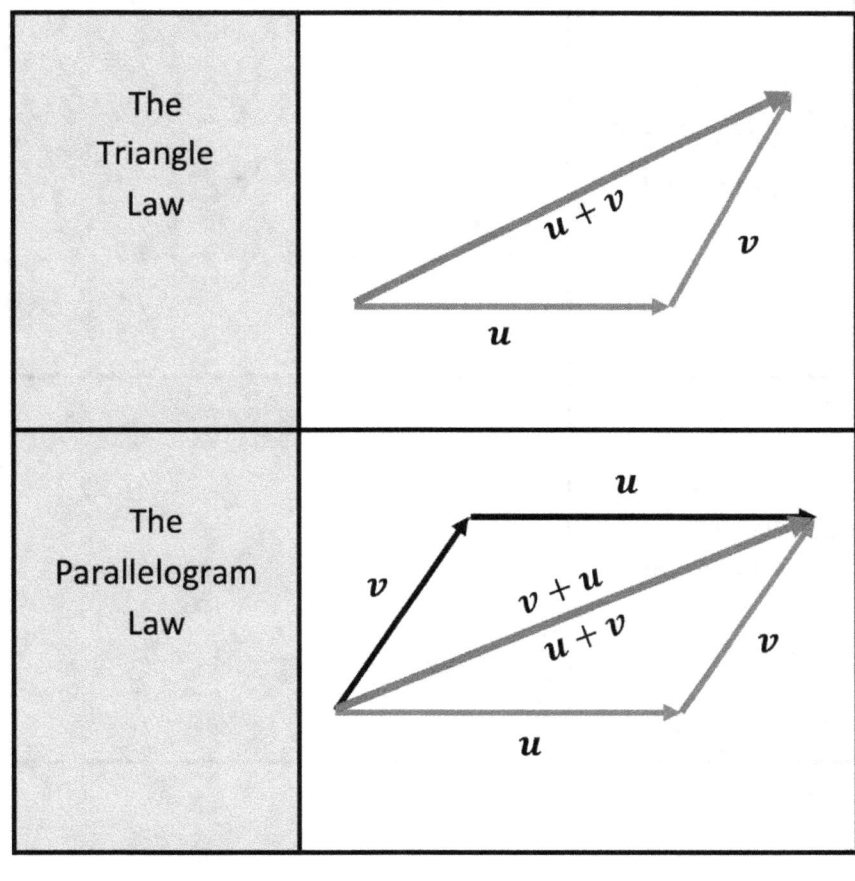
The Parallelogram Law	

Vector Subtraction -- $\mathbb{R}^2$ Illustration

Two vectors, u and v

Subtracted "tail to tail"

Place vectors "tail to tail"	
Place vectors "tail to tail"	

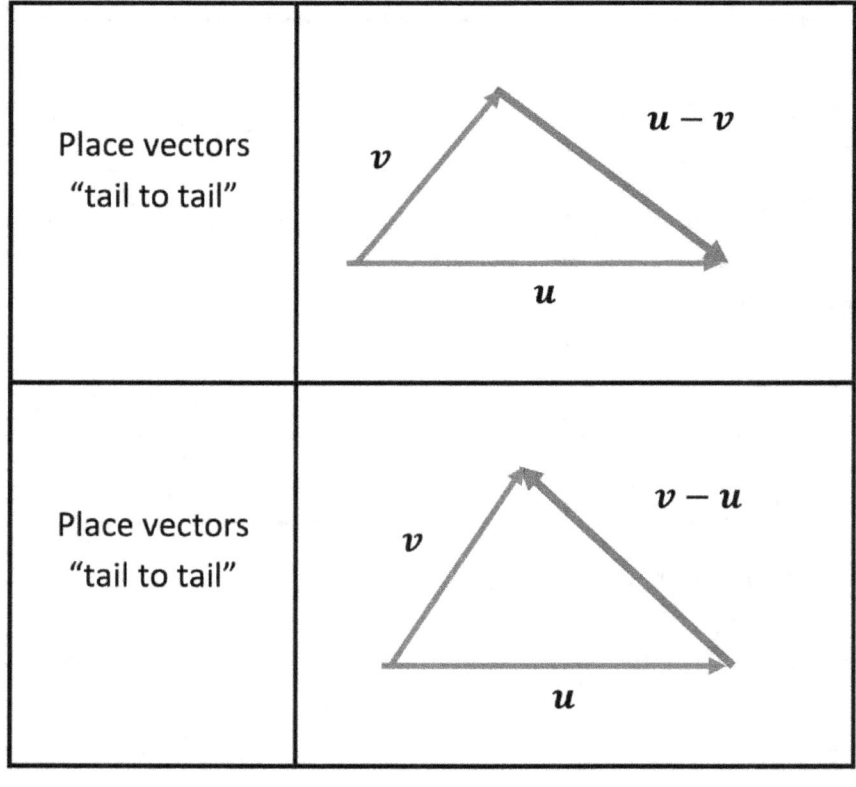

Vectors -- Scalar Multiplication	

Given a constant, c (or scalar) and vector, v

Then ...

$$\text{The Scalar Multiple} = cv$$

Length of cv	$|cv| = $ magnitude of cv $\|cv\| = \|c\| \cdot \|v\|$	
Direction of cv	If $c > 0$ then c is positive. The direction of cv is the <u>same</u> as v	
	If $c < 0$ then c is negative. The direction of cv is the <u>opposite</u> of v	
	If $c = 0$ or $v = 0$ Then $cv = 0$ (no direction)	

Vectors -- Scalar Multiplication -- Algebra

Given: vector, $v = \langle x_1, y_1, z_1 \rangle$ and constant, c

Then:

$cv = $ The scalar multiple

$cv = c \langle x_1, y_1, z_1 \rangle$

$cv = \langle c\,x_1,\ c\,y_1,\ c\,z_1 \rangle$

Vectors -- Scalar Multiplication -- Example

Given: vector, $v = \langle 1,\ 2,\ -3 \rangle$

And a constant, $c = 5$

Find the scalar multiple: cv

$cv = 5 \langle 1,\ 2, -3 \rangle$

$cv = \langle 5,\ 10,\ -15 \rangle$

Vectors -- Scalar Multiplication -- Illustrated

v		$-v$	
$2v$		$-2v$	
$\frac{1}{2}v$		$-\frac{1}{2}v$	

Vectors -- From Point to Point

Given the following points:

$$O = (\,0, 0, 0\,)$$

$$A = (\,a_1, a_2, a_3\,)$$

$$B = (\,b_1, b_2, b_3\,)$$

$\overrightarrow{OA}$	= Position vector of A = $\langle\, a_1, a_2, a_3\,\rangle$
$\overrightarrow{OB}$	= Position vector of B = $\langle\, b_1, b_2, b_3\,\rangle$
$\overrightarrow{AB}$	= Vector from A to B = $\langle\, b_1 - a_1, b_2 - a_2, b_3 - a_3\,\rangle$
$\overrightarrow{BA}$	= Vector from B to A = $\langle\, a_1 - b_1, a_2 - b_2, a_3 - b_3\,\rangle$

Vectors -- From Point to Point -- Example

Find vectors $\overrightarrow{AB}$ and $\overrightarrow{BA}$

Given the following points:

$A = (1, 2, 3)$ and $B = (8, 5, -6)$

$\overrightarrow{AB}$	$= $ Vector from A to B $= \langle b_1 - a_1, b_2 - a_2, b_3 - a_3 \rangle$ $= \langle 8 - 1, 5 - 2, -6 - 3 \rangle$ $= \langle 7, 3, -9 \rangle$
$\overrightarrow{BA}$	$= $ Vector from B to A $= \langle a_1 - b_1, a_2 - b_2, a_3 - b_3 \rangle$ $= \langle 1 - 8, 2 - 5, 3 - (-6) \rangle$ $= \langle -7, -3, 9 \rangle$

Vectors -- Magnitude

The magnitude of a vector is its length and is defined as follows:

If: $\quad v = \langle a, b, c \rangle$

Then: $\quad |v| =$ The magnitude of v

$$|v| = \sqrt{a^2 + b^2 + c^2}$$

Vectors -- Magnitude -- Example

Given: $v = \langle 1, 2, -3 \rangle$

Find the magnitude of v

$$|v| = \sqrt{1^2 + 2^2 + (-3)^2}$$

$$|v| = \sqrt{1 + 4 + 9}$$

$$|v| = \sqrt{14}$$

Vectors -- 2D and 3D

A 2D vector in the xy plane is represented

In $\mathbb{R}^2$ as: $\boldsymbol{v} = \langle\, a, b \,\rangle$

That same vector can be represented as a 3D vector

In $\mathbb{R}^3$ as: $\boldsymbol{v} = \langle\, a, b, 0 \,\rangle$

Note:

The xy plane is where $z = 0$ in $\mathbb{R}^3$

Vectors -- Properties

Vectors: u, v, w Scalars: a, b, c

$$u + v = v + u$$

$$u + (v + w) = (u + v) + w$$

$$u + 0 = u$$

$$u + (-u) = 0$$

$$c(u + v) = cu + cv$$

$$(a + b)u = au + bu$$

$$(ab)u = a(bu)$$

$$1u = u$$

Unit Vectors

A unit vector has a length of 1.

Given some vector, a , with length $|a|$

Then:

u = Unit vector in the direction of a

$$u = \frac{a}{|a|}$$

Standard Basis Vectors

$$i = \langle 1,0,0 \rangle \qquad j = \langle 0,1,0 \rangle \qquad k = \langle 0,0,1 \rangle$$

Standard basis vectors are unit vectors

in the direction of positive x, y and z axes.

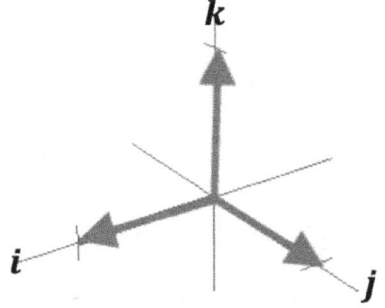

Vectors -- Ex. 1
Given: $\quad\quad a = \langle\, 2,0,1\,\rangle \quad\quad b = \langle\, -2,1,3\,\rangle$ Find the following ...

$\lvert a \rvert$	$= \sqrt{2^2 + 0^2 + 1^2} \;\;=\;\; \sqrt{5}$
$a + b$	$= \langle\, 2 - 2, 0 + 1, 1 + 3\,\rangle$ $= \langle\, 0\,,1\,,\,4\,\rangle$
$a - b$	$= \langle\, 2 + 2, 0 - 1, 1 - 3\,\rangle$ $= \langle\, 4\,,-1\,,-2\,\rangle$
$3b$	$= \langle\, -6\,,\,3\,,\,9\,\rangle$
$2a + 4b$	$= 2\langle\, 2,0,1\,\rangle \,+\, 4\langle -2,1,3\,\rangle$ $= \langle\, 4,0,2\,\rangle \,+\, \langle -8,4,12\,\rangle$ $= \langle\, 4 - 8,\; 0 + 4,\; 2 + 12\,\rangle$ $= \langle\, -4,\; 4,\; 14\,\rangle$

Vectors -- Ex. 2

Express the vector $2a + 3b$

in terms of i, j, k

Given:

$$a = \langle 2, 0, 1 \rangle$$
$$b = \langle -2, 1, 3 \rangle$$

$2a + 3b$

$$= 2\langle 2, 0, 1 \rangle + 3\langle -2, 1, 3 \rangle$$

$$= \langle 4, 0, 2 \rangle + \langle -6, 3, 9 \rangle$$

$$= \langle -2, 3, 11 \rangle$$

$$= -2i + 3j + 11k$$

Vectors -- Ex. 3

Find the unit vector in the direction of

$$w = 2i - 6j + 3k$$

$|w|$	$|w|$ = Magnitude of vector w $|w| = \sqrt{2^2 + (-6)^2 + 3^2}$ $|w| = \sqrt{4 + 36 + 9}$ $|w| = \sqrt{49} = 7$
u	u = Unit vector in direction of w $u = \dfrac{w}{|w|}$ $u = \dfrac{1}{7}\langle 2, -6, 3 \rangle$ $u = \langle \dfrac{2}{7}, -\dfrac{6}{7}, \dfrac{3}{7} \rangle$ $u = \dfrac{2}{7}i - \dfrac{6}{7}j + \dfrac{3}{7}k$

Vectors -- Ex. 4a

A 200 pound person hangs from 2 wires.
Find the tensions (forces) T_1 and T_2
and the magnitude of the tensions.

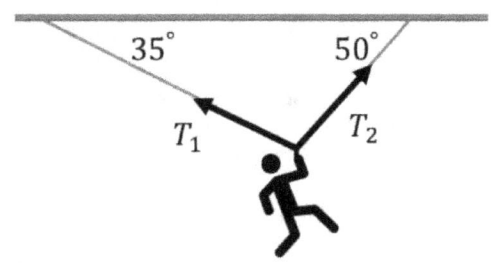

T_1	$= \|T_1\| \langle -\cos 35, \sin 35 \rangle$
T_2	$= \|T_2\| \langle \cos 50, \sin 50 \rangle$
Horiz. Forces	$\|T_1\|(\cos 35) \quad = \quad \|T_2\|(\cos 50)$ $\dfrac{\|T_1\| \cos 35}{\cos 50} \quad = \quad \|T_2\|$
Vert. Forces	$\|T_1\|(\sin 35) + \|T_2\|(\sin 50) \quad = \quad 200$
$\|T_1\|(\sin 35) + \dfrac{\|T_1\| \cos 35}{\cos 50}(\sin 50) \quad = \quad 200$	

Vectors -- Ex. 4b

A 200 pound person hangs from 2 wires.
Find the tensions (forces) T_1 and T_2
and the magnitude of the tensions.

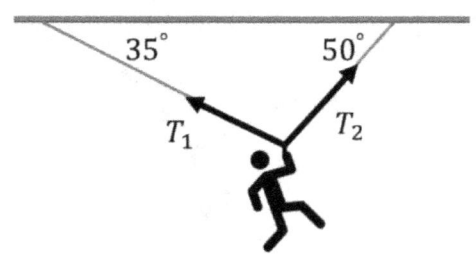

Find the magnitude of the tensions, $|T_1|$ and $|T_2|$

$$|T_1| (\sin 35) + \frac{|T_1| \cos 35}{\cos 50} (\sin 50) = 200$$

$$|T_1| \left[(\sin 35) + \frac{(\cos 35)}{\cos 50} (\sin 50) \right] = 200$$

$$|T_1| [\, 1.5498\,] \approx 200$$

$$|T_1| \approx \frac{200}{1.5498} \approx 129.05$$

$$|T_2| = \frac{|T_1|(\cos 35)}{\cos 50} \approx \frac{129.05\,(\cos 35)}{\cos 50}$$

$$|T_2| \approx 164.46$$

Vectors -- Ex. 4c

A 200 pound person hangs from 2 wires.
Find the tensions (forces) T_1 and T_2
and the magnitude of the tensions.

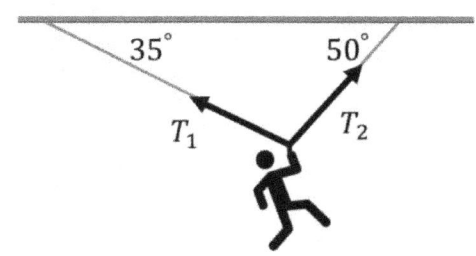

Find the tensions (forces), T_1 and T_2

$$T_1 = |T_1| \langle -\cos 35, \sin 35 \rangle$$

$$T_1 = 129.05 \langle -\cos 35, \sin 35 \rangle$$

$$T_1 = \langle -105.71, 74.02 \rangle$$

$$T_2 = |T_2| \langle \cos 50, \sin 50 \rangle$$

$$T_2 = 164.46 \langle \cos 50, \sin 50 \rangle$$

$$T_2 = 164.46 \langle 105.71, 125.98 \rangle$$

The Dot Product

Dot Product

The Dot Product is used to find the angle between two vectors. The good news is that it is easy to calculate!

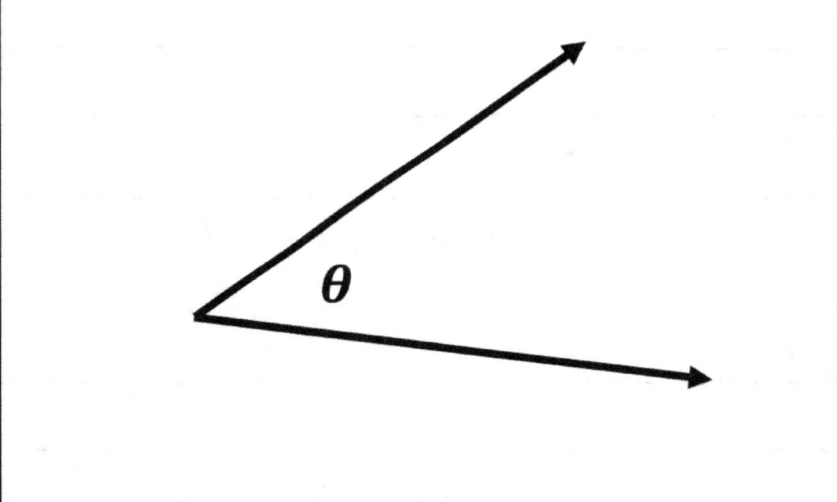

Dot Product

Used to find the angle between 2 vectors and is easy to calculate. Given:

$$\boldsymbol{u} = \langle\, a, b, c \,\rangle \quad \text{and} \quad \boldsymbol{v} = \langle\, d, e, f \,\rangle$$

Dot Product -- Calculation

$$\boldsymbol{u} \cdot \boldsymbol{v} \;=\; \text{Dot Product}$$

$$\boldsymbol{u} \cdot \boldsymbol{v} \;=\; ad + be + cf$$

Dot Product and θ

$$\boldsymbol{u} \cdot \boldsymbol{v} \;=\; |\boldsymbol{u}|\,|\boldsymbol{v}| \cos\theta$$

$$\cos\theta \;=\; \frac{\boldsymbol{u} \cdot \boldsymbol{v}}{|\boldsymbol{u}|\,|\boldsymbol{v}|}$$

$$\theta \;=\; \cos^{-1}\left(\frac{\boldsymbol{u} \cdot \boldsymbol{v}}{|\boldsymbol{u}|\,|\boldsymbol{v}|} \right)$$

Dot Product

$$a \cdot b \;=\; |a|\,|b|\cos\theta$$

If $\cos\theta = 0$ Then $\theta = 90°$

And the Dot Product $=\; a \cdot b \;=\; 0$

So the two vectors are orthogonal.

Dot Product -- More Properties

$a \cdot a$	$=$	$|a|^2$
$a \cdot b$	$=$	$b \cdot a$
$a \cdot (b + c)$	$=$	$a \cdot b \;+\; a \cdot c$
$(ca) \cdot b$	$=$	$c(a \cdot b) \;=\; a \cdot (cb)$
$0 \cdot a$	$=$	0

Dot Product -- Ex. 1

Find the angle between two vectors.
Given:

$$u = \langle 1, 2, 3 \rangle \quad \text{and} \quad v = \langle 4, 0, -1 \rangle$$

$$u \cdot v = \langle 1, 2, 3 \rangle \cdot \langle 4, 0, -1 \rangle$$

$$u \cdot v = (1)(4) + (2)(0) + (3)(-1)$$

$$u \cdot v = 4 + 0 - 3 = 1$$

$$|u| = \sqrt{1^2 + 2^2 + 3^2} = \sqrt{14}$$

$$|v| = \sqrt{4^2 + 0^2 + (-1)^2} = \sqrt{17}$$

$$u \cdot v = |u| \, |v| \cos \theta$$

$$\cos \theta = \frac{u \cdot v}{|u| \, |v|}$$

$$\theta = \cos^{-1}\left(\frac{1}{\sqrt{14} \, \sqrt{17}} \right) \approx 86.3°$$

Dot Product -- Ex. 2

Show the two vectors are perpendicular.

Given:

$$u = \langle\, 4, 4, -2\,\rangle \quad \text{and} \quad v = \langle\, 5, -4, 2\,\rangle$$

$$u \cdot v = \langle\, 4, 4, -2\,\rangle \cdot \langle\, 5, -4, 2\,\rangle$$

$$u \cdot v = (4)(5) + (4)(-4) + (-2)(2)$$

$$u \cdot v = 20 - 16 - 4 = 0$$

If the Dot Product is zero,

then the vectors are orthogonal (perpendicular)

because: $\cos \theta = \cos 90° = 0$

$$u \cdot v = |u|\,|v| \cos \theta$$

$$\cos \theta = \frac{u \cdot v}{|u|\,|v|}$$

$$\theta = \cos^{-1}\left(\frac{0}{\sqrt{14}\,\sqrt{17}}\right) \approx 90°$$

Direction Angles and Projections

Direction Angles

The Direction Angles of a vector a are:

α = Angle between a and the x-axis

β = Angle between a and the y-axis

γ = Angle between a and the z-axis

Unit Vector

The unit vector, in the direction of a is:

$$\frac{a}{|a|} = \langle \cos\alpha, \cos\beta, \cos\gamma \rangle$$

Direction Cosines

Components of the unit vector. For a

$\cos\alpha$ = Component in x-direction.

$\cos\beta$ = Component in y-direction.

$\cos\gamma$ = Component in z-direction.

Direction Angles -- Example	
Find the direction angles for vector a . Given: $a = \langle 1, 2, 5 \rangle$	

Magnitude Of vector a	$\|a\| = \sqrt{1^2 + 2^2 + 5^2}$ $\|a\| = \sqrt{1 + 4 + 25} = \sqrt{30}$
Cosines of the direction angles.	$\cos\alpha = \dfrac{1}{\sqrt{30}}$ $\cos\beta = \dfrac{2}{\sqrt{30}}$ $\cos\gamma = \dfrac{5}{\sqrt{30}}$
Direction Angles	$\alpha = \cos^{-1}\left(\dfrac{1}{\sqrt{30}}\right) \approx 80°$ $\beta = \cos^{-1}\left(\dfrac{2}{\sqrt{30}}\right) \approx 69°$ $\gamma = \cos^{-1}\left(\dfrac{5}{\sqrt{30}}\right) \approx 24°$

	Projection of b onto a	

Scalar projection of b onto a	$comp_a\, b$	$= \dfrac{a \cdot b}{\lvert a \rvert}$
Vector projection of b onto a	$proj_a\, b$	$= \left(\dfrac{a \cdot b}{\lvert a \rvert} \right) \dfrac{a}{\lvert a \rvert}$ $= \dfrac{a \cdot b}{\lvert a \rvert^2}\, a$

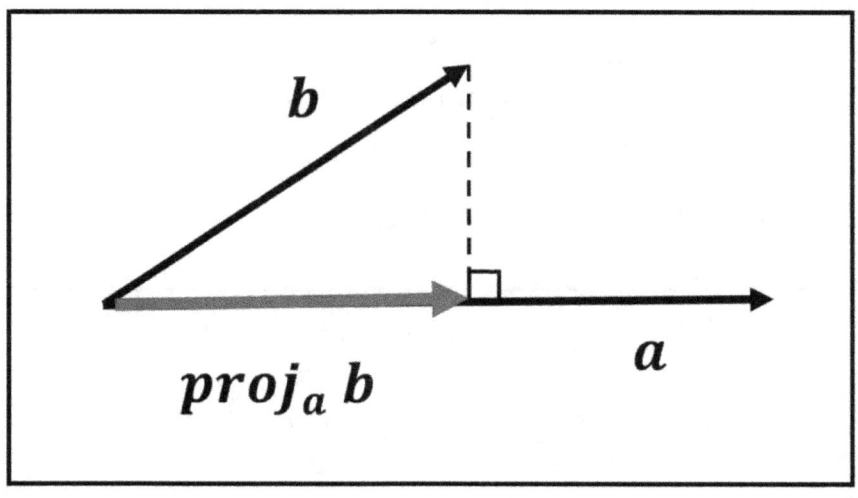

Projections -- Example

Find the scalar and vector projections

of vector b onto vector a.

Given: $a = \langle\, 1, -2, 5\,\rangle$ $b = \langle\, 0, 1, 3\,\rangle$

$a \cdot b = \langle\, 1, -2, 5\,\rangle \cdot \langle\, 0, 1, 3\,\rangle$

$a \cdot b = (1)(0) + (-2)(1) + (5)(3)$

$a \cdot b = 13$

$|a| = \sqrt{1^2 + (-2)^2 + 5^2} = \sqrt{30}$

$comp_{\,a}\, b$ = Scalar projection of b onto a

$comp_{\,a}\, b = \dfrac{a \cdot b}{|a|} = \dfrac{13}{\sqrt{30}}$

$proj_{\,a}\, b$ = Vector projection of b onto a

$proj_{\,a}\, b = \left(\dfrac{a \cdot b}{|a|}\right) \dfrac{a}{|a|} = \dfrac{a \cdot b}{|a|^2}\, a$

$proj_{\,a}\, b = \dfrac{13}{30} \langle\, 1, -2, 5\,\rangle$

Direction Angles -- Work

W = Work = FD = Force x Distance

$W = (F \cos \theta)D$

$W = |F||D| \cos \theta$

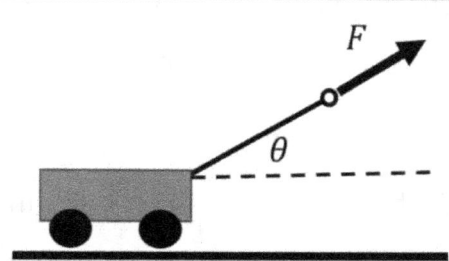

Direction Angles -- Work Example

In the diagram, above, find Work done.

Given: $F = 50\,N$, $D = 20\,m$, $\theta = 30°$

$W = |F||D| \cos \theta$

$W = (50)(20) \cos 30$

$W = (50)(20)\left(\frac{\sqrt{3}}{2}\right) \approx 866\,J$

Work -- From Point A to Point B

Find the word done when a force moves a particle from point A to point B.

Given: $A = (2, 1, 0)$ $B = (5, 6, 1)$

The force vector is: $F = \langle 1, 2, 3 \rangle$

Distance is in meters, Force in in N.

$\overrightarrow{AB}$ = Displacement Vector

$\overrightarrow{AB} = \langle 5 - 2, \ 6 - 1, \ 1 - 0 \rangle$

$\overrightarrow{AB} = \langle 3, 5, 1 \rangle$

$W = F \cdot D$

$W = \langle 1, 2, 3 \rangle \cdot \langle 3, 5, 1 \rangle$

$W = (1)(3) + (2)(5) + (3)(1)$

$W = 3 + 10 + 3 = 16J$

The Cross Product

Cross Product

Used to find a 3rd vector that is perpendicular

to two other vectors.   Note: The result is a **VECTOR**.

Cross Product -- Calculation

Not so easy to calculate. A little tedious.

Explained after the determinant review

on the next page.

Cross Product -- Illustration

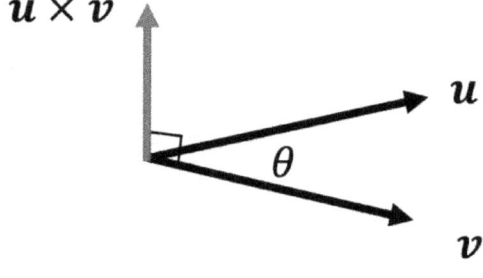

Determinant Review

A determinant of order 2 is defined by:

$$\begin{vmatrix} a & b \\ c & d \end{vmatrix} = ad - bc$$

Multiply the diagonals and subtract.

A determinant of order 3 is defined by:

$$\begin{vmatrix} a_1 & a_2 & a_3 \\ b_1 & b_2 & b_3 \\ c_1 & c_2 & c_3 \end{vmatrix} =$$

$$+ \; a_1 \begin{vmatrix} b_2 & b_3 \\ c_2 & c_3 \end{vmatrix}$$

$$- \; a_2 \begin{vmatrix} b_1 & b_3 \\ c_1 & c_3 \end{vmatrix}$$

$$+ \; a_3 \begin{vmatrix} b_1 & b_2 \\ c_1 & c_2 \end{vmatrix}$$

Cross Product -- Calculation

The cross product of two vectors is a 3rd vector that is orthogonal to both vectors. Given two vectors:

$$a = \langle a_1, a_2, a_3 \rangle \qquad b = \langle b_1, b_2, b_3 \rangle$$

$$a \times b = \text{The Cross Product}$$

Calculated below ...

$a \times b =$

$$\begin{vmatrix} i & j & k \\ a_1 & a_2 & a_3 \\ b_1 & b_2 & b_3 \end{vmatrix} =$$

$$+ i \begin{vmatrix} a_2 & a_3 \\ b_2 & b_3 \end{vmatrix}$$

$$- j \begin{vmatrix} a_1 & a_3 \\ b_1 & b_3 \end{vmatrix}$$

$$+ k \begin{vmatrix} a_1 & a_2 \\ b_1 & b_2 \end{vmatrix}$$

Cross Product -- Properties
Vector $a \times b$ is orthogonal to a and b
If θ is the angle between a and b $\| a \times b \| = \|a\| \|b\| \sin \theta$
Two non-zero vectors are parallel IFF $a \times b = 0$

$a \times b$	$=$	$-b \times a$
$(c\,a) \times b$	$=$	$c(a \times b) = a \times (cb)$
$a \times (b + c)$	$=$	$a \times b + a \times c$
$(a + b) \times c$	$=$	$a \times c + b \times c$
$a \cdot (b \times c)$	$=$	$(a \times b) \cdot c$
$a \times (b \times c)$	$=$	$(a \cdot c)b - (a \cdot b)c$

Scalar Triple Product
Volume of a Parallelepiped

The scalar triple product of 3 vectors

$$a = \langle a_1, a_2, a_3 \rangle$$
$$b = \langle b_1, b_2, b_3 \rangle$$
$$c = \langle c_1, c_2, c_3 \rangle$$

Is calculated as a determinant.

$$a \cdot (b \times c) = \begin{vmatrix} a_1 & a_2 & a_3 \\ b_1 & b_2 & b_3 \\ c_1 & c_2 & c_3 \end{vmatrix}$$

The volume of a **parallelepiped** is the magnitude of the scalar triple product.

$$V = |a \cdot (b \times c)|$$

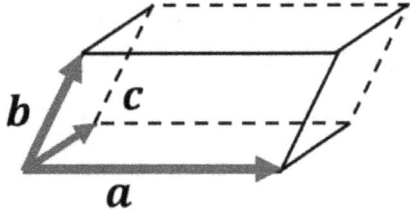

Cross Product -- Ex. 1a

Find the Cross Product $\boldsymbol{a} \times \boldsymbol{b}$

Given: 2 vectors $\quad \boldsymbol{a} = \langle 1, 2, 3 \rangle \quad \boldsymbol{b} = \langle 5, -1, 0 \rangle$

$$\begin{vmatrix} i & j & k \\ 1 & 2 & 3 \\ 5 & -1 & 0 \end{vmatrix} =$$

$$+\ i \begin{vmatrix} 2 & 3 \\ -1 & 0 \end{vmatrix}$$

$$-\ j \begin{vmatrix} 1 & 3 \\ 5 & 0 \end{vmatrix}$$

$$+\ k \begin{vmatrix} 1 & 2 \\ 5 & -1 \end{vmatrix}$$

VISUAL HINTS ...

$$\begin{vmatrix} i & j & k \\ 1 & 2 & 3 \\ 5 & -1 & 0 \end{vmatrix}$$

$$\begin{vmatrix} i & j & k \\ 1 & 2 & 3 \\ 5 & -1 & 0 \end{vmatrix}$$

$$\begin{vmatrix} i & j & k \\ 1 & 2 & 3 \\ 5 & -1 & 0 \end{vmatrix}$$

Continued ...

Cross Product -- Ex. 1b
Find the Cross Product $a \times b$ Given: 2 vectors $\quad a = \langle 1, 2, 3 \rangle \quad\quad b = \langle 5, -1, 0 \rangle$

$+i \begin{vmatrix} 2 & 3 \\ -1 & 0 \end{vmatrix}$	$= i \,[(2)(0) - (3)(-1)]$ $= i \,[0 + 3\,]$ $= \,i \,(3)$
$-j \begin{vmatrix} 1 & 3 \\ 5 & 0 \end{vmatrix}$	$= -j \,[\,(1)(0) - (3)(5)]$ $= -j \,[\,0 - 15]$ $= \,-j \,(\,15)$
$+k \begin{vmatrix} 1 & 2 \\ 5 & -1 \end{vmatrix}$	$= k \,[(1)(-1) - (2)(5)]$ $= k \,[\,-1 - 10\,]$ $= k \,(-11)$
$a \times b \;=\; \langle\, 3, -15, -11 \,\rangle$	

Equations of Lines

Vector Equation of a Line

$$L = r_0 + tv$$

The Vector Equation of a Line can be found if given a point on a line and a direction vector for the line.

$$P_0 = (x_0, y_0, z_0) \quad \text{and} \quad v = \langle a, b, c \rangle$$

Use the position vector of the point and the direction vector: $\quad r_0 = \langle x_0, y_0, z_0 \rangle \quad$ and $\quad v = \langle a, b, c \rangle$

$$L = r_0 + tv$$
$$L = \langle x_0, y_0, z_0 \rangle + t\langle a, b, c \rangle$$
$$L = \langle x_0 + ta, \ y_0 + tb, \ z_0 + c \rangle$$

Parametric Equations of a Line

Parametric Equations for a line through point (x_0, y_0, z_0) and parallel to direction vector $\langle a, b, c \rangle$	$x = x_0 + ta$ $y = y_0 + tb$ $z = z_0 + tc$

	Vector Equation of a Line -- Example $$L = r_0 + tv$$

Find the vector and parametric equations for the line that passes through the point $(4, 1, 2)$ and is parallel to the vector $\langle 1, 3, -2 \rangle$.

Also, find 2 points on the line.

Vector Equation	$L = r_0 + tv$ $L = \langle 4, 1, 2 \rangle + t \langle 1, 3, -2 \rangle$ $L = \langle 4 + t, 1 + 3t, 2 - 2t \rangle$
Parametric Equations	$x = 4 + t$ $y = 1 + 3t$ $z = 2 - 2t$
2 Points (x, y, z)	$t = 1 \quad \rightarrow \quad (5, 4, 0)$ $t = 2 \quad \rightarrow \quad (6, 7, -2)$

Symmetric Equation of a Line

The Symmetric Equation of a Line through a point

$P_0 = (x_0, y_0, z_0)$ and parallel to $v = \langle a, b, c \rangle$

is:

$$\frac{x - x_0}{a} = \frac{y - y_0}{b} = \frac{z - z_0}{c}$$

If $a = 0$, then

$$x = x_0, \quad \frac{y - y_0}{b} = \frac{z - z_0}{c}$$

NOTE: The parameter " t " is eliminated by setting all parametric equations $= t$ and, therefore, equal to each other.

Recall, the Parametric Equations are:

$$x = x_0 + ta$$
$$y = y_0 + tb$$
$$z = z_0 + tc$$

Symmetric Equation of a Line -- Ex. 1a

Find the Symmetric and Parametric Equations for the line through 2 points. $A\,(1, 2, -1)$ and $B\,(3, 1, 2)$.

Where does it intersect the xy plane?

Direction vector	$\boldsymbol{v} = \overrightarrow{AB}$ $\boldsymbol{v} = \langle\, 3 - 1, 1 - 2, 2 + 1\,\rangle$ $\boldsymbol{v} = \langle\, 2, -1, 3\,\rangle$
Parametric Eqns.	$x\ =\ \ 1 + 2t$ $y\ =\ \ 2 + t$ $z\ =\ -1 + 3t$
Symmetric Eqns.	$\dfrac{x-1}{2} = \dfrac{y-2}{-1} = \dfrac{z+1}{3}$
xy plane Intersect. Set $z = 0$	$\dfrac{x-1}{2} = \dfrac{y-2}{-1} = \dfrac{1}{3}$ $x = \dfrac{5}{3},\ \ y = \dfrac{5}{3},\ \ z = 0$

Symmetric Equation of a Line -- Ex. 1b (Extra)

Find the Symmetric and Parametric Equations for the line through 2 points. $A\ (1, 2, -1)$ and $B\ (3, 1, 2)$.

Where does it intersect the xy plane?

Previously Found: Parametric Eqns.	$x = 1 + 2t$ $y = 2 + t$ $z = -1 + 3t$
Note	$t = 0 \quad \rightarrow \quad$ Point A $t = 1 \quad \rightarrow \quad$ Point B
Line Segment from point A to point B.	$r = \langle 1 + 2t, 2 + t, -1 + 3t \rangle$ $0 \leq t \leq 1$

Line Segment from r_0 to r_1	$r(t) = (1 - t)r_0 + t\,r_1$ $0 \leq t \leq 1$

Skew Lines -- Ex. 2

Show that two lines are skew.

(Do not intersect and are not parallel.)

Given parametric equations for the lines.

L_1: $x = 1 + t$	L_2: $x = 2s$
$y = -2 + 3t$	$y = 3 + s$
$z = 4 - t$	$z = -3 + 4s$

Not parallel because the direction vectors

$\langle 1, 3, -1 \rangle$ and $\langle 2, 1, 4 \rangle$ are not parallel.

Do not intersect because no values of t & s

will make the components equal.

$x = 1 + t = 2s$ $\rightarrow$ $t = 2s - 1$

$y = -2 + 3t = 3 + s$ $\rightarrow$ $t = \dfrac{5 + s}{3}$

$z = 4 - t = -3 + 4s$ $\rightarrow$ $t = 7 - 4s$

Substituting $t = (2s - 1)$

gives different values of s for the other eqns.

Stewart, Calculus Early Transcendentals, p. 826

Equations of Planes

Scalar Equation of a Plane

A plane in space is determined by a point in the plane and a vector that is normal to the plane.

The Scalar Equation of a plane can be found if given a point in the plane and a vector, normal to the plane.

$$P_0 = (x_0, y_0, z_0) \quad \text{and} \quad \boldsymbol{n} = \langle a, b, c \rangle$$

$$a(x - x_0) + b(y - y_0) + c(z - z_0) = 0$$

Proof

Point $\rightarrow$ Position Vectors

$$P_0 = (x_0, y_0, z_0) \quad \rightarrow \quad \boldsymbol{r_0} = \langle x_0, y_0, z_0 \rangle$$
$$P = (x, y, z) \quad \rightarrow \quad \boldsymbol{r} = \langle x, y, z \rangle$$

The normal vector is perpendicular to the vector $\langle r - r_0 \rangle$ so the Dot Product $= 0$

$$\boldsymbol{n} \cdot \langle r - r_0 \rangle = 0$$
$$\langle a, b, c \rangle \cdot \langle x - x_0, y - y_0, z - z_0 \rangle = 0$$
$$a(x - x_0) + b(y - y_0) + c(z - z_0) = 0$$

Distance From a Point to a Plane

$$D \; = \; \frac{|\, a\, x_1 \; + \; b y_1 \; + \; c\, z_1 \; + \; d \,|}{\sqrt{a^2 + b^2 + c^2}}$$

Plane Equation: $\qquad ax + by + cz + d \; = \; 0$

Vector $\perp$ to plane: $\quad \boldsymbol{n} \; = \; \langle\, a, b, c \,\rangle$

Point in the plane: $\qquad P_0 \; = \; (\, x_0, y_0, z_0 \,)$

Point not in plane: $\qquad P_1 \; = \; (\, x_1, y_1, z_1 \,)$

Let: $\boldsymbol{b} \; = \; \overrightarrow{P_0\, P_1}$

$\qquad \boldsymbol{b} \; = \; \langle\, x_1 - x_0, \; y_1 - y_0, \; z_1 - z_0 \,\rangle$

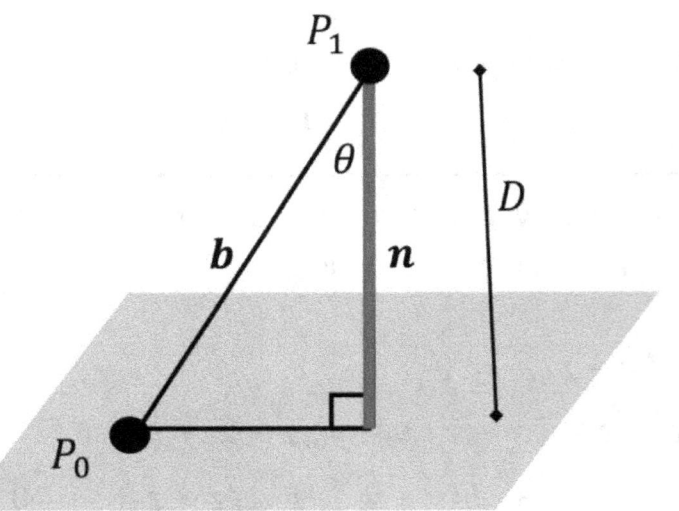

Distance From a Point to a Plane -- Proof

$$D \;=\; |\, comp_n b \,| \;=\; \frac{|\, n \cdot b \,|}{|\, n \,|}$$

$$D \;=\; \frac{|\, a(x_1 - x_0) + b(y_1 - y_0) + c\,(z_1 - x_0) \,|}{\sqrt{a^2 + b^2 + c^2}}$$

$$D \;=\; \frac{(\, ax_1 + by_1 + cz_1 \,)}{\sqrt{a^2 + b^2 + c^2}}$$

$$-\; \frac{(\, ax_0 + by_0 + cz_0 + d - d \,)}{\sqrt{a^2 + b^2 + c^2}}$$

Recall: $ax_0 + by_0 + cz_0 + d \;=\; 0$

$$D \;=\; \frac{|\, a\,x_1 + by_1 + c\,z_1 + d \,|}{\sqrt{a^2 + b^2 + c^2}}$$

Plane Equation: $ax + by + cz + d \;=\; 0$

Vector $\perp$ to plane: $n \;=\; \langle\, a, b, c \,\rangle$

Point in the plane: $P_0 \;=\; (\, x_0, y_0, z_0 \,)$

Point not in plane: $P_1 \;=\; (\, x_1, y_1, z_1 \,)$

Let: $b \;=\; \overrightarrow{P_0 P_1}$

$\quad\quad b \;=\; \langle\, x_1 - x_0,\; y_1 - y_0,\; z_1 - z_0 \,\rangle$

Scalar Equation of a Plane -- Ex. 1a

Find Scalar Equation of a plane given a point in plane and n, normal to plane.

$$P_0 = (1, 5, -2) \quad \text{and} \quad n = \langle 2, 3, 4 \rangle$$

Also, find intercepts and sketch the plane

$$a(x - x_0) + b(y - y_0) + c(z - z_0) = 0$$

$$2(x - 1) + 3(y - 5) + 4(z + 2) = 0$$

$$2x - 2 + 3y - 15 + 4z + 8 = 0$$

$$2x + 3y + 4z = 9$$

x-intercept $y = z = 0$	$2x = 9 \quad \rightarrow \quad x = \frac{9}{2}$ Point: $\left(\frac{9}{2}, 0, 0\right)$
y-intercept $x = z = 0$	$3y = 9 \quad \rightarrow \quad y = 3$ Point: $(0, 3, 0)$
z-intercept $x = y = 0$	$4z = 9 \quad \rightarrow \quad x = \frac{9}{4}$ Point: $\left(0, 0, \frac{9}{4}\right)$

Scalar Equation of a Plane -- Ex. 1b

Find Scalar Equation of a plane given a point in plane and n, normal to plane.

$$P_0 = (1, 5, -2) \quad \text{and} \quad n = \langle 2, 3, 4 \rangle$$

Also, find intercepts and sketch the plane

Previously Found	$2x + 3y + 4z = 9$
Intercepts	$\left(\frac{9}{2}, 0, 0\right), (0, 3, 0), \left(0, 0, \frac{9}{4}\right)$
Linear Equation Format	$ax + by + cz - d = 0$ $2x + 3y + 4z - 9 = 0$ With: $n = \langle a, b, c \rangle$
Sketch	

Scalar Equation of a Plane -- Ex. 2a

Find Scalar Equation of a plane given 3 points on the plane.

$$P(\,1,2,3\,) \qquad Q(\,3,-1,5\,) \qquad R(\,4,2,0\,)$$

Also, sketch the plane.

$\overrightarrow{PQ}$	$\overrightarrow{PQ} = \langle\, 3-1, -1-2, 5-3\,\rangle$ $\overrightarrow{PQ} = \langle\, 2, -3, 2\,\rangle$
$\overrightarrow{PR}$	$\overrightarrow{PR} = \langle\, 4-1, 2-2, 0-3\,\rangle$ $\overrightarrow{PR} = \langle\, 3, 0, -3\,\rangle$
n	$n = \overrightarrow{PQ} \times \overrightarrow{PR}$ $n = \langle\, 2, -3, 2\,\rangle \times \langle\, 3, 0, -3\,\rangle$ $n = \begin{vmatrix} i & j & k \\ 2 & -3 & 2 \\ 3 & 0 & -3 \end{vmatrix}$ $n = (9)i - (-12)j + (9)k$ $n = \langle\, 9, 12, 9\,\rangle = 3\langle\, 3, 4, 3\,\rangle$

Scalar Equation of a Plane -- Ex. 2b

Find Scalar Equation of a plane given 3 points on the plane.

$$P(1,2,3) \qquad Q(3,-1,5) \qquad R(4,2,0)$$

Also, sketch the plane.

Previously Found: $\quad \boldsymbol{n} = \langle 3,4,3 \rangle$

Scalar Equation of a Plane

With $\quad P(1,2,3) \quad$ and $\quad \boldsymbol{n} = \langle 3,4,3 \rangle$

$$3(x-1) + 4(y-2) + 3(z-3) = 0$$
$$3x - 3 + 4y - 8 + 3z - 9 = 0$$
$$3x + 4y + 3z = 20$$

Sketch	
Use given points	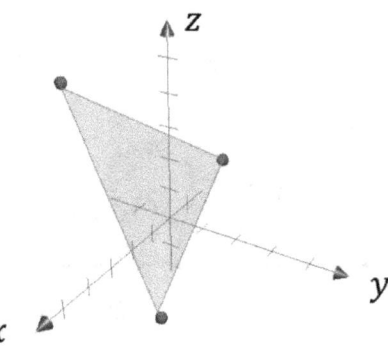

Scalar Equation of a Plane -- Ex. 3a

Find the angle between 2 planes. Also, find line of

intersection. Given 2 planes:

$$x + 2y + z = 1 \qquad \text{and} \qquad x - y + 4z = 1$$

Find Normal vectors	$n_1 = \langle 1, 2, 1 \rangle$ $n_2 = \langle 1, -1, 4 \rangle$

Angle between normal vectors
= Angle between the planes.

$$\cos \theta \;=\; \frac{n_1 \cdot n_2}{|\, n_1 \,|\,|\, n_2 \,|}$$

$$=\; \frac{(1)(1) \;+\; (2)(-1) \;+\; (1)(4)}{\sqrt{1^2 + 2^2 + 1^2}\;\sqrt{1^2 + (-1)^2 + 4^2}}$$

$$=\; \frac{1 - 2 + 4}{\sqrt{6}\;\sqrt{18}} \;=\; \frac{3}{\sqrt{108}}$$

$$\theta \;=\; \cos^{-1}\left(\frac{3}{\sqrt{108}}\right) \;\approx\; 73^{\circ}$$

Continued ...

Scalar Equation of a Plane -- Ex. 3b

Find the angle between 2 planes. Also, find line of

intersection. Given 2 planes:

$$x + 2y + z = 1 \qquad \text{and} \qquad x - y + 4z = 1$$

Previously Found	$n_1 = \langle 1, 2, 1 \rangle$
	$n_2 = \langle 1, -1, 4 \rangle$

Find a point on line of intersection:

Let $z = 0$ (where line crosses xy-plane)

Solve both equations for x and equate.

$$
\begin{array}{ccc}
1 - 2y & = & 1 + y \\
0 & = & 3y \\
0 & = & y
\end{array}
$$

$x = 1 - 2(0) \qquad \rightarrow \qquad x = 1$

Point on line of intersection: $(1, 0, 0)$

Continued ...

Scalar Equation of a Plane -- Ex. 3c

Find the angle between 2 planes. Also, find line of intersection. Given 2 planes:

$$x + 2y + z = 1 \qquad \text{and} \qquad x - y + 4z = 1$$

Previously Found	$n_1 = \langle 1, 2, 1 \rangle$
	$n_2 = \langle 1, -1, 4 \rangle$
	Point on line: $(1, 0, 0)$

Direction vector of the line is orthogonal to the normal vectors of the 2 planes.

$$v = n_1 \times n_2 = \begin{vmatrix} i & j & k \\ 1 & 2 & 1 \\ 1 & -1 & 4 \end{vmatrix}$$

$$v = \langle 9, -3, -3 \rangle$$

Symmetric Equations of the Line:

$$\frac{x - 1}{9} = \frac{y - 0}{-3} = \frac{z - 0}{-3}$$

Distance Between Parallel Planes -- Ex. 4a

Find the distance (D) between two parallel planes:

$$2x + 3y - z = 4 \qquad \text{and}$$

$$4x + 6y - 2z = 1$$

Note: The planes are parallel because their normal vectors are parallel.

$$n_1 = \langle\, 2, 3, -1 \,\rangle \qquad n_2 = \langle\, 4, 6, -2 \,\rangle$$

To find the distance, find a point on either plane then calculate the distance to the other plane.

$$\text{Set:} \quad y = z = 0$$

1ˢᵗ Plane	$2x = 4 \qquad \rightarrow \qquad x = 2$ $P_1 = (\,2, 0, 0\,)$
2ⁿᵈ Plane *Only need one point.*	$4x = 1 \qquad \rightarrow \qquad x = \dfrac{1}{4}$ $P_2 = \left(\dfrac{1}{4}, 0, 0\right)$

Continued ...

Distance Between Parallel Planes -- Ex. 4b

Find the distance (D) between two parallel planes:

$$2x + 3y - z = 4 \quad \text{and}$$

$$4x + 6y - 2z = 1$$

Previously Found 2 Points	$P_1 = (2,0,0)$
	$P_2 = \left(\frac{1}{4}, 0, 0\right)$

Use distance formula to find distance from point $P_1 = (2,0,0)$ to the other plane.

Equation of 2nd Plane: $\quad 4x + 6y - 2z - 1 = 0$

$$n = \langle 2, 3, -1 \rangle \quad \text{or} \quad n = \langle 4, 6, -2 \rangle$$

$$D = \frac{|a x_1 + b y_1 + c z_1 + d|}{\sqrt{a^2 + b^2 + c^2}}$$

$$D = \frac{|2(2) + 3(0) - 1(0) - 1|}{\sqrt{2^2 + 3^2 + (-1)^2}}$$

$$D = \frac{|4 - 1|}{\sqrt{14}} = \frac{3}{\sqrt{14}}$$

Surfaces – Cylinders & Quadrics

Surfaces -- Cylinders & Quadratics
In this section, equations for two types of 3D surfaces are reviewed – Cylinders and Quadratics.

Equations for Cylinders	Include 2 of the 3 variables (x, y, z).
Equations for Quadratics	Include 3 of the 3 variables (x, y, z) and at least one of the variables is squared.

Surfaces -- Equations of Cylinders

A cylinder is a surface consisting of all lines (rulings) parallel to a given line and pass through a given plane curve. Some examples are shown below.

$y = x^2$ Parallel to z-axis	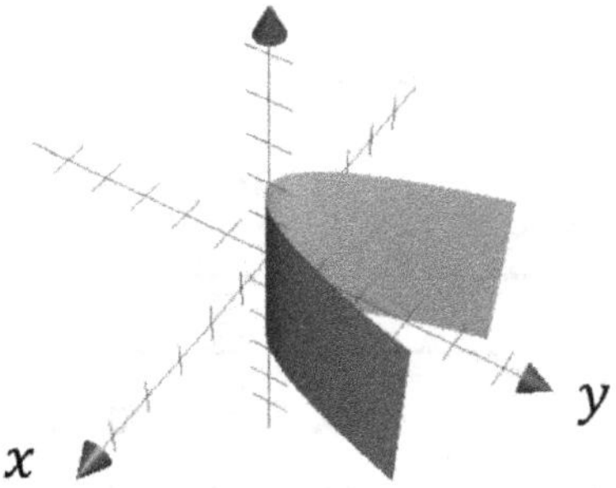
$x^2 + y^2 = 1$ Parallel to z-axis	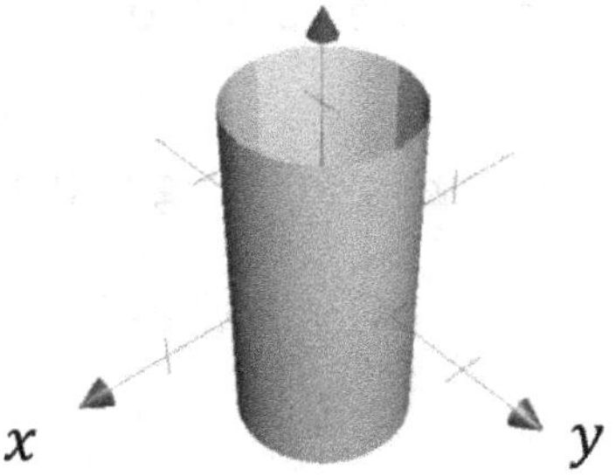
$x^2 + z^2 = 1$ Parallel to y-axis	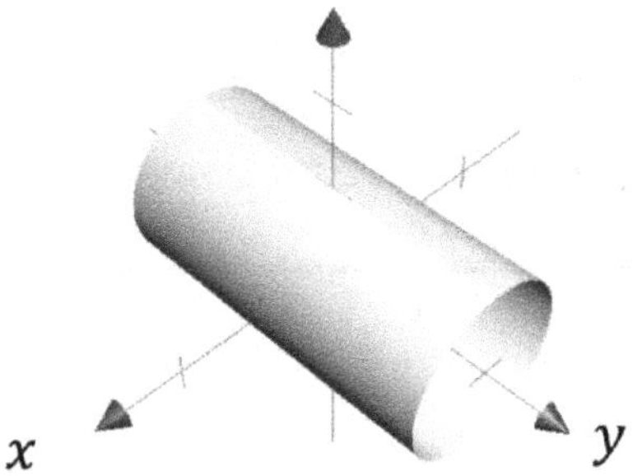

Surfaces -- Equations of Quadratics

A quadric surface is a 2nd degree equation, involving 3 variables: x, y, & z.

Quadrics can be put into the form :

$$Ax^2 + By^2 + Cz^2$$
$$+ Dxy + Exz + Fyz$$
$$+ Gx + Hy + Iz + J = 0$$

To sketch quadrics, use <u>Traces</u>.

Set one of the variables to a constant, k

Set $z = k$ to get traces in the xy Plane

Set $y = k$ to get traces in the xz Plane

Set $x = k$ to get traces in the yz Plane

For each plane, sketch traces for various values of k.

Sketching several traces on one graph is often helpful.

Surfaces -- Quadratic Surfaces (Part 1)

Ellipsoid $$\frac{x^2}{a^2} + \frac{y^2}{b^2} + \frac{z^2}{c^2} = 1$$	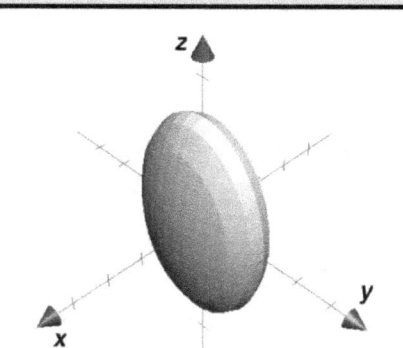
Cone $$\frac{x^2}{a^2} + \frac{y^2}{b^2} = \frac{z^2}{c^2}$$	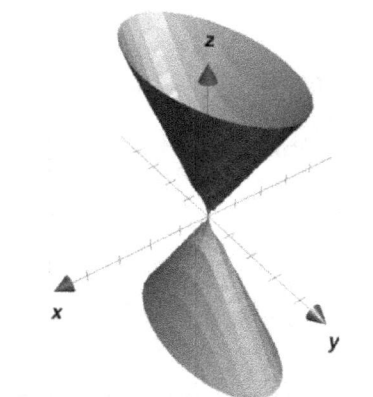
Elliptic Paraboloid $$\frac{x^2}{a^2} + \frac{y^2}{b^2} = \frac{z}{c}$$	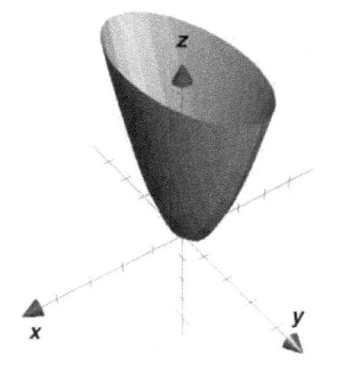

Surfaces -- Quadratic Surfaces (Part 2)

Hyperboloid Of One Sheet $$\frac{x^2}{a^2} + \frac{y^2}{b^2} - \frac{z^2}{c^2} = 1$$	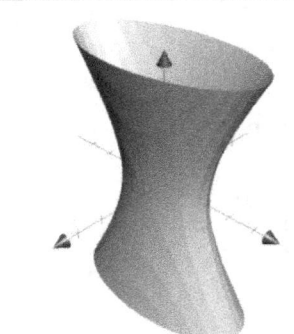
Hyperbolic Paraboloid $$\frac{x^2}{a^2} - \frac{y^2}{b^2} = \frac{z}{c}$$	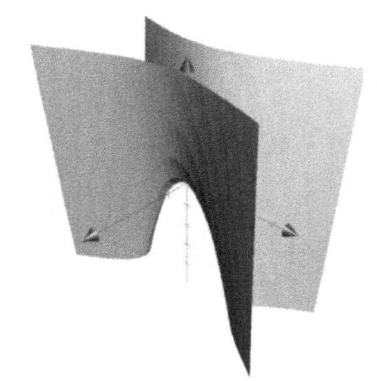
Hyperboloid Of Two Sheets $$-\frac{x^2}{a^2} - \frac{y^2}{b^2} + \frac{z^2}{c^2} = 1$$	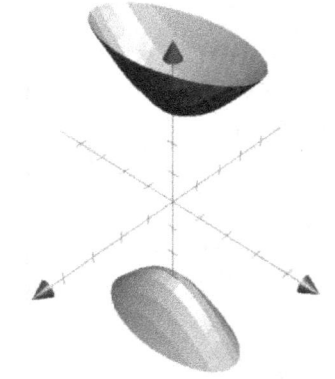

Surfaces -- Quadratic Surfaces -- Ex. 1a

Sketch: $x^2 + \dfrac{y^2}{9} + \dfrac{z^2}{4} = 1$

Set $z = k$ $x^2 + \dfrac{y^2}{9} = 1 - \dfrac{k^2}{4}$ Ellipse: $k \leq 2$ Traces: $k = 0, 1$	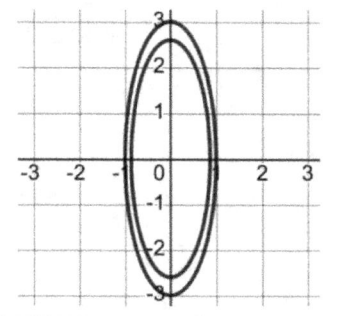
Set $y = k$ $x^2 + \dfrac{z^2}{4} = 1 - \dfrac{k^2}{9}$ Ellipse: $k \leq 3$ Traces: $k = 0, 2$	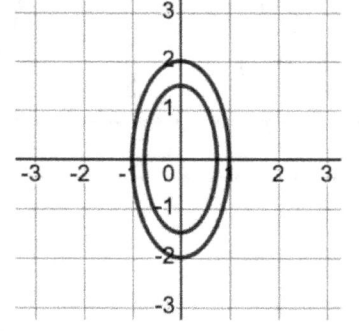
Set $x = k$ $\dfrac{y^2}{9} + \dfrac{z^2}{4} = 1 - k^2$ Ellipse: $k \leq 1$ Traces: $k = 0, \dfrac{1}{2}$	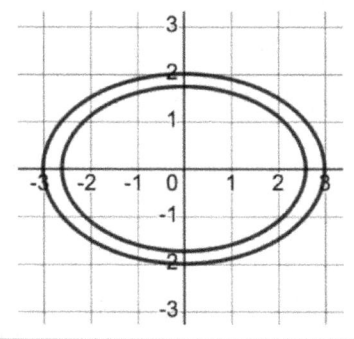

Surfaces -- Quadratic Surfaces -- Ex. 1b

Sketch: $\quad x^2 + \dfrac{y^2}{9} + \dfrac{z^2}{4} = 1$

Previously found traces ...

$z = k$ xy Plane	$y = k$ xz Plane	$x = k$ yz Plane
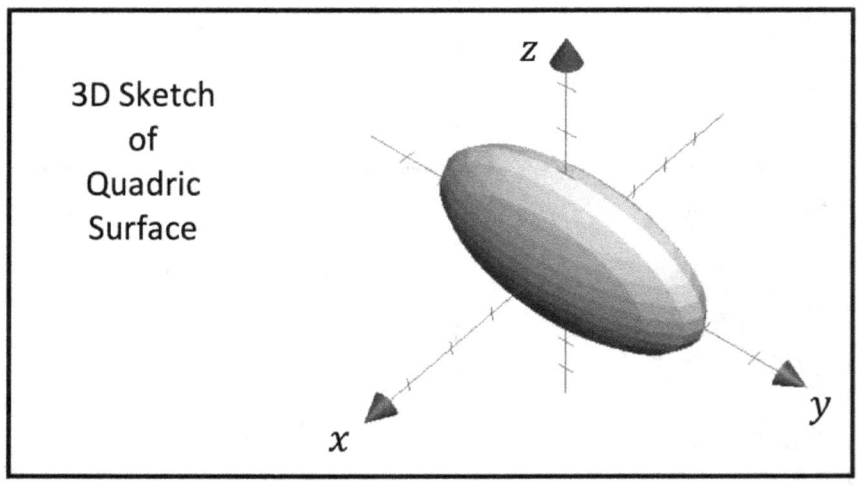		

3D Sketch of Quadric Surface

Surfaces -- Quadratic Surfaces -- Ex. 2a

Sketch: $4x^2 - y^2 + 2z^2 + 4 = 0$

Rearrange the given equation

to help identify it.

$$4x^2 - y^2 + 2z^2 + 4 = 0$$

$$x^2 - \frac{y^2}{4} + \frac{z^2}{2} + 1 = 0$$

$$-x^2 + \frac{y^2}{4} - \frac{z^2}{2} = 1$$

It is a Hyperboloid of 2 Sheets.

Continued ...

Surfaces -- Quadratic Surfaces -- Ex. 2b

Use this equation: $-x^2 + \dfrac{y^2}{4} - \dfrac{z^2}{2} = 1$

Set $z = k$ $-x^2 + \dfrac{y^2}{4} = 1 + \dfrac{k^2}{2}$ Hyperbola: Traces: $k = 0,\ 2$	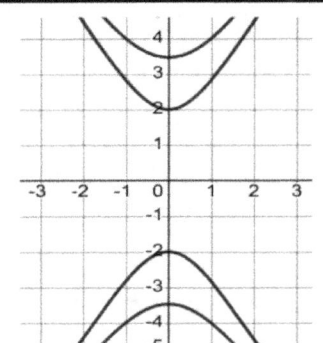
Set $y = k$ $x^2 + \dfrac{z^2}{2} = \dfrac{k^2}{4} - 1$ Ellipse: $\lvert k \rvert \geq 2$ Traces: $k = \pm 3,\ \pm 4$	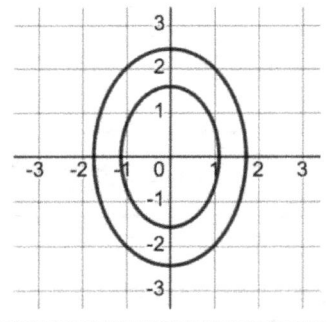
Set $x = k$ $\dfrac{y^2}{4} - \dfrac{z^2}{2} = 1 + k^2$ Hyperbola: Traces: $k = 0,\ 1$	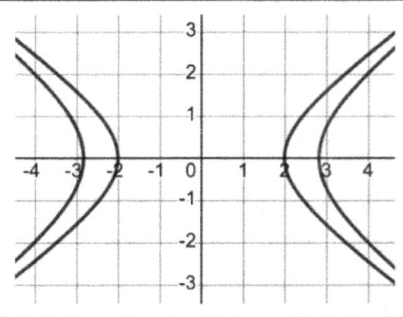

Surfaces -- Quadratic Surfaces -- Ex. 2c

Use this equation: $\quad -x^2 + \dfrac{y^2}{4} - \dfrac{z^2}{2} = 1$

Previously found traces ...

$z = k$	$y = k$	$x = k$
xy Plane	xz Plane	yz Plane

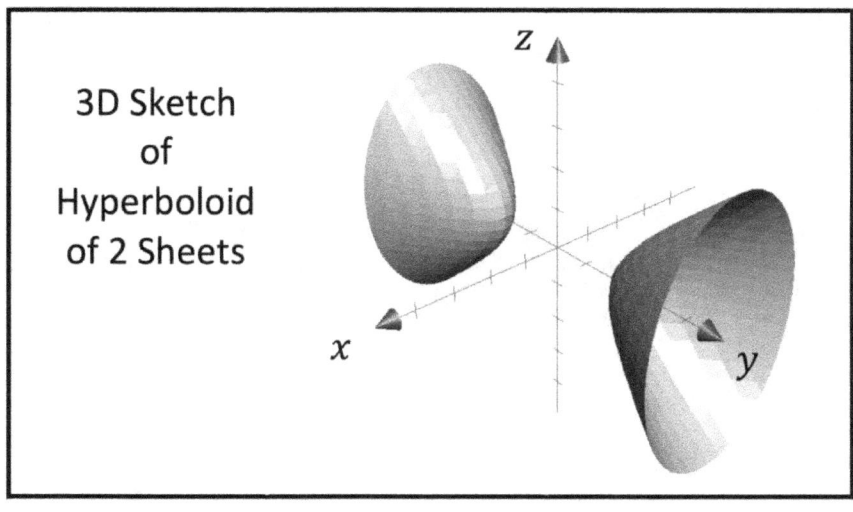

3D Sketch
of
Hyperboloid
of 2 Sheets

Vector Functions

Vector Functions and Space Curves

Vector Functions

A vector function is a function whose

<u>domain</u> is the set of real numbers and whose

<u>range</u> is a set of vectors.

$r(t)$ = 3D Vector function

$r(t)$ = $\langle f(t), g(t), z(t) \rangle$

$r(t)$ = $f(t)\,\boldsymbol{i} + g(t)\,\boldsymbol{j} + z(t)\,\boldsymbol{k}$

Where:

$f(t), g(t), h(t)$ = Component functions

t = Independent variable (usually time)

Space Curves

A **Space Curve** is the path of a particle along a vector function, for a range of t values (or time interval).

Vector Functions -- Limits

If $\quad \boldsymbol{r(t)} \; = \; \langle \, f(t), \; g(t), \; z(t) \, \rangle \qquad$ Then ...

$$\lim_{t \to a} \boldsymbol{r(t)} \; = \; \langle \, \lim_{t \to a} f(t), \; \lim_{t \to a} f(t) \, , \; \lim_{t \to a} f(t) \, \rangle$$

Provided the limits of the component functions exist.

A vector function, $\boldsymbol{r}$, is continuous at a if

$$\lim_{t \to a} \boldsymbol{r(t)} \; = \; \boldsymbol{r}(a)$$

Also: A vector function, $\boldsymbol{r}$, is continuous at a

IFF its component functions are continuous at a .

Vector Functions -- Limits -- Ex. 1

Find: $\lim\limits_{t \to 0} r(t)$

Where: $r(t) = \langle 1 + t^2,\ 2e^{-t},\ \dfrac{\sin t}{t} \rangle$

$\lim\limits_{t \to 0} (1 + t^2)$	$= (1 + 0) = 1$
$\lim\limits_{t \to 0} (2e^{-t})$	$= (2)(1) = 2$
$\lim\limits_{t \to 0} \left(\dfrac{\sin t}{t} \right)$	$= 1$
$\lim\limits_{t \to 0} r(t) = \langle 1, 2, 1 \rangle$ $\lim\limits_{t \to 0} r(t) = i + 2j + k$	

Vector Functions -- Space Curve -- Ex. 2

Describe the curve defined by:

$$r(t) = \langle\, 3 + t, 2 - t, 1 + 3t \,\rangle$$

Parametric equations of a line, passing through point $(3, 2, 1)$ and parallel to vector $\langle\, 1, -1, \ 3 \,\rangle$	$x = 3 + t$ $y = 2 - t$ $z = 1 + 3t$
Sketch	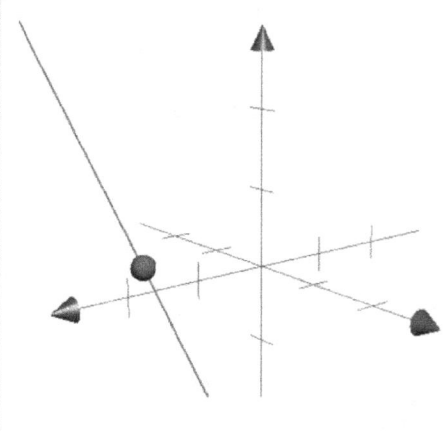

Vector Functions -- Space Curve -- Ex. 3

Describe the curve defined by:

$$r(t) = \langle \cos t, \sin t, t \rangle$$

The curve lies within a circular cylinder. The curve spirals upward, around the cylinder as t increases.	$x = \cos t$ $y = \sin t$ $z = t$
	$\cos^2 t + \sin^2 t = 1$ $x^2 + y^2 = 1$
Sketch Helix	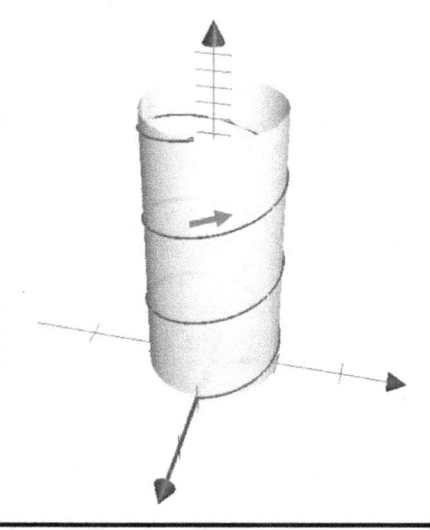

Vector Functions -- Space Curve -- Ex. 4

Find a vector equation and parametric equations

for the line segment that joins the points

$P(1, 0, 2)$ and $Q(3, 4, 5)$

Eqn. that joins tips of vectors r_0 & r_1

$$r(t) = (1 - t)\, r_0 + t\, r_1 \qquad 0 \le t \le 1$$

$$r(t) = (1 - t)\langle 1, 0, 2 \rangle + t \langle 3, 4, 5 \rangle$$

Vector Equations:

$$i = (1 - t)1 + 3t = 1 + 2t$$

$$j = (1 - t)0 + 4t = 4t$$

$$k = (1 - t)2 + 5t = 2 + 3t$$

$$r(t) = \langle 1 + 2t,\ 4t,\ 2 + 3t \rangle , \qquad 0 \le t \le 1$$

Parametric Equations: $\quad 0 \le t \le 1$

$$x = 1 + 2t \ , \quad y = 4t \ , \quad z = 2 + 3t$$

Vector Functions -- Space Curve -- Ex. 5

Find a vector equation that represents the curve of

intersection of the cylinder: $x^2 + y^2 = 1$

and the plane: $x + z = 3$

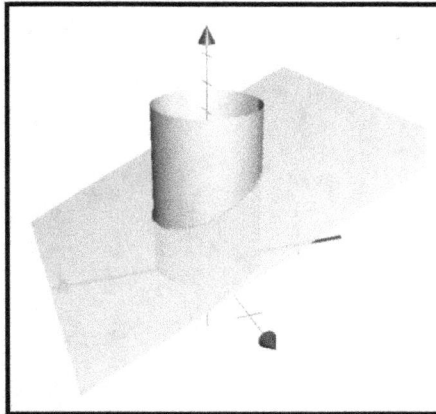 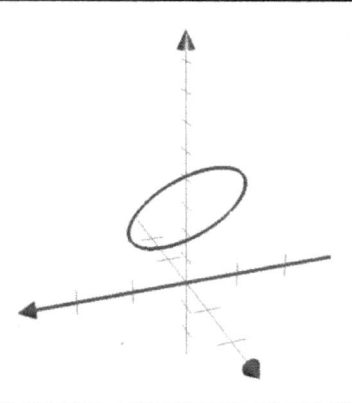

The projection of the curve onto the xy plane

is the circle: $x^2 + y^2 = 1$

So: $x = \cos t$, $y = \sin t$ With: $0 \le t \le 2\pi$

And: $z = 3 - x = 3 - \cos t$

Vector Equation	$r(t) = \langle \cos t, \sin t, 3 - \cos t \rangle$
	$0 \le t \le 2\pi$

Derivatives of Vector Functions

Derivatives of Vector Functions
$r'(t) \; = \; \frac{d}{dt}\, r(t)$
$r(t) \quad = \quad \langle\, f(t),\; g(t),\; h(t)\,\rangle$
$r'(t) \quad = \quad \langle\, f'(t),\; g'(t),\; h'(t)\,\rangle$
$r'(t) \quad = \quad f'(t)\, i + \; g'(t)\, j + \; h'(t)\, k$

Derivatives of Vector Functions -- Rules

$$u, v = \text{Differentiable vectors}$$
$$f = \text{Real-valued function}$$

$$\frac{d}{dt}\left[u(t) + v(t) \right] = u'(t) + v'(t)$$

$$\frac{d}{dt}\left[c\, u(t) \right] = c\, u'(t)$$

$$\frac{d}{dt}\left[f(t)\, u(t) \right] = f'\, u + f\, u'$$

$$\frac{d}{dt}\left[u(t) \cdot v(t) \right] = u' \cdot v + u \cdot v'$$

$$\frac{d}{dt}\left[u(t) \times v(t) \right] = u' \times v + u \times v'$$

$$\frac{d}{dt}\left[u(f(t)) \right] = f'\, u'(f(t)) \qquad \text{Chain Rule}$$

Derivatives of Vector Functions -- Ex. 1

Show that if $|\mathbf{r}(t)| = c$

Then $r'(t)$ is orthogonal to $r(t)$ for all t

$|r(t)|$	$=$	c	Given
$|r(t)|^2$	$=$	c^2	
$r(t) \cdot r(t)$	$=$	c^2	

$$\frac{d}{dt}[r(t) \cdot r(t)] = \frac{d}{dt}[c^2] = 0$$

$$\frac{d}{dt}[r(t) \cdot r(t)] = r' \cdot r + r \cdot r'$$

$$0 = 2\,r' \cdot r$$

$$0 = r' \cdot r$$

Since $r' \cdot r = 0$ (Dot Product $= 0$)

$r'(t)$ is orthogonal to $r(t)$

(Stewart, Calculus Early Transcendentals, p. 858)

Integrals of Vector Functions

Integrals of Vector Functions

$$\boldsymbol{r(t)} = \langle f(t),\ g(t),\ h(t) \rangle$$

$$\int \boldsymbol{r(t)}\,dt = \langle \int f\,dt,\ \int g\,dt,\ \int h\,dt \rangle$$

$$\int \boldsymbol{r(t)}\,dt = \int f\,dt\,\boldsymbol{i} + \int g\,dt\,\boldsymbol{j} + \int h\,dt\,\boldsymbol{k}$$

$$\int_a^b r(t)\,dt = R(b) - R(a)$$

Where:

R = antiderivative of r

$R' = r$

Integrals of Vector Functions -- Ex. 1

Find: $\int r(t)\, dt$

Given: $r(t) = \langle\, \cos t, 3\sin t, 2t\,\rangle$

$I = Integral$

$I = \int r(t)\, dt$

$I = \langle\, \sin t, -3\cos t, \frac{2}{2}t^2\,\rangle + C$

$I = \sin t\, \boldsymbol{i} - 3\cos t\, \boldsymbol{j} + t^2 \boldsymbol{k} + C$

Integrals of Vector Functions -- Ex. 2

Find: $\int_0^{\frac{\pi}{3}} r(t)\, dt$

Given: $r(t) = \langle \cos t, 3\sin t, 2t \rangle$

$I = \int_0^{\frac{\pi}{3}} r(t)\, dt$

$I = [\,\sin t\, i - 3\cos t\, j + t^2 k\,]_0^{\frac{\pi}{3}}$

$I = \left[\, \sin\left(\frac{\pi}{3}\right) i - 3\cos\left(\frac{\pi}{3}\right) j + \left(\frac{\pi}{3}\right)^2 k \,\right]$

$\qquad\qquad - [\,\sin(0)\, i - 3\cos(0)\, j + (0)^2 k\,]$

$I = \left[\, \frac{\sqrt{3}}{2} i - 3\left(\frac{1}{2}\right) j + \left(\frac{\pi}{3}\right)^2 k \,\right]$

$\qquad\qquad - [\,(0)i - 3(1)j + (0)^2 k\,]$

$I = \langle\, \left(\frac{\sqrt{3}}{2} - 0\right), \left(-\frac{3}{2} + \frac{6}{2}\right), \left(\frac{\pi^2}{9} - 0\right) \,\rangle$

$I = \langle\, \frac{\sqrt{3}}{2}, \frac{3}{2}, \frac{\pi^2}{9} \,\rangle$

Arc Length

Arc Length of Vector Functions

Given: $r(t) = \langle f(t), g(t), h(t) \rangle$

L = Arc Length

$$L = \int_a^b \sqrt{[f'(t)]^2 + [g'(t)]^2 + [h'(t)]^2}\ dt$$

$$L = \int_a^b \sqrt{\left(\frac{dx}{dt}\right)^2 + \left(\frac{dy}{dt}\right)^2 + \left(\frac{dz}{dt}\right)^2}\ dt$$

$$L = \int_a^b |r'(t)|\ dt$$

Arc Length of Vector Functions -- Ex. 1

Find the length of the arc of a circular helix, for the

given vector: $r(t) = \langle \cos t, \sin t, t \rangle$

From point $(1, 0, 0)$ to point $(1, 0, 4\pi)$

$r'(t) = \langle -\sin t, \cos t, 1 \rangle$

$|r'(t)| = \sqrt{(-\sin t)^2 + \cos^2 t + 1} = \sqrt{2}$

Starting at $t = 0$ \qquad Ending at $t = 4\pi$

$L = $ Length of Arc

$L = \int_a^b |r'(t)| \, dt$

$L = \int_0^{4\pi} \sqrt{2} \, dt$

$L = \sqrt{2} \, [t]_0^{4\pi}$

$L = \sqrt{2} \, [4\pi - 0]$

$L = 4\pi \sqrt{2}$

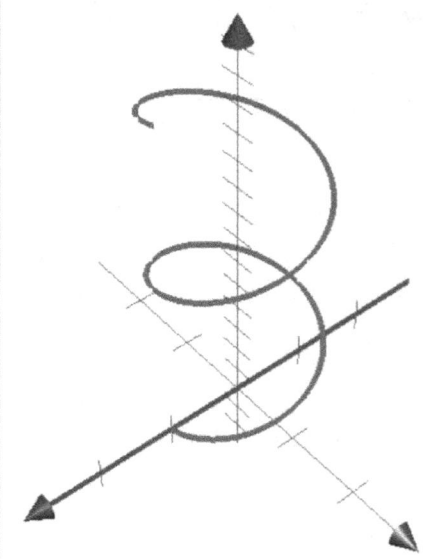

Curvature

Curvature of Vector Functions	
Given:	$r(t) = \langle f(t),\ g(t),\ h(t) \rangle$

Unit Tangent Vector	$T(t) = \dfrac{r'(t)}{\lvert r'(t) \rvert}$
Curvature of vector $r(t)$	$k(t) = \dfrac{\lvert T'(t) \rvert}{\lvert r'(t) \rvert}$ $k(t) = \dfrac{\lvert r'(t) \times r''(t) \rvert}{\lvert r'(t) \rvert^3}$
Curvature of Plane curve $y = f(x)$	$k(x) = \dfrac{\lvert f''(x) \rvert}{\left[1 + (f'(x))^2 \right]^{3/2}}$

Curvature of Vector Functions -- Ex. 1

Find the curvature of the twisted cubic

$$r(t) = \langle t, t^2, t^3 \rangle \quad \text{at point } (0,0,0)$$

$r'(t)$	$r'(t) = \langle 1, 2t, 3t^2 \rangle$
$r''(t)$	$r''(t) = \langle 0, 2, 6t \rangle$
$\lvert r'(t) \rvert$	$\lvert r'(t) \rvert = \sqrt{1 + 4t^2 + 9t^4}$
$r'(t) \times r''(t) = \begin{vmatrix} i & j & k \\ 1 & 2t & 3t^2 \\ 0 & 2 & 6t \end{vmatrix} = \langle 6t^2, -6t, 2 \rangle$	

$$\lvert r'(t) \times r''(t) \rvert = \sqrt{36t^4 + 36t^2 + 4}$$
$$= 2\sqrt{9t^4 + 9t^2 + 1}$$

$$k(t) = \frac{\lvert r'(t) \times r''(t) \rvert}{\lvert r'(t) \rvert^3}$$

$$k(t) = \frac{2\sqrt{9t^4 + 9t^2 + 1}}{(1 + 4t^2 + 9t^4)^{3/2}}$$

$$k(0) = \frac{2\sqrt{1}}{(1)^{3/2}} = \frac{2}{1} = 2 \qquad t = 0$$

(Stewart, Calculus Early Transcendentals, p. 865)

Curvature of Vector Functions -- Ex. 2

Find curvature of the parabola $y = z^2$

at points: $(y, z) = (0, 0)$ and $(0, 1)$

$f(z)$	$f(z) = z^2$
$f'(z)$	$f'(z) = 2z$
$f''(z)$	$f''(z) = 2$
$k(z)$ Curvature	$k(z) = \dfrac{\lvert f''(z) \rvert}{[\, 1 \; + \; (f'(z))^2\,]^{3/2}}$ $k(z) = \dfrac{\lvert 2 \rvert}{[\, 1 + 4z^2\,]^{3/2}}$
At point $(y, z) = (0, 0)$	$k(0) = 2$
At point $(y, z) = (0, 1)$	$k(1) = \dfrac{\lvert 2 \rvert}{[1 + 4]^{\frac{3}{2}}}$ $k(1) = \dfrac{2}{5\sqrt{5}} \; \approx \; 0.18$

Normal, Binormal and Tangent Vectors

Normal, Binormal, and Tangent Unit Vectors
Given: $r(t) = \langle f(t), g(t), h(t) \rangle$

Unit Tangent Vector	$T(t) = \dfrac{r'(t)}{\lvert r'(t) \rvert}$
Unit Normal Vector	$N(t) = \dfrac{T'(t)}{\lvert T'(t) \rvert}$
Unit Binormal Vector	$B(t) = T(t) \times N(t)$
Curvature	$k = \dfrac{\lvert T'(t) \rvert}{\lvert r'(t) \rvert}$ $k = \dfrac{\lvert r'(t) \times r''(t) \rvert}{\lvert r'(t) \rvert^3}$
B, N, T Unit Vectors	

Normal, Binormal, & Tangent Unit Vectors -- Ex. 1

Find Normal, Binormal, and Tangent unit vectors for:

$r(t) = \langle \cos t, \sin t, t \rangle$ Curve is a circular helix.

$r'(t) = \langle -\sin t, \cos t, 1 \rangle$

$|r'(t)| = \sqrt{\sin^2 t + \cos^2 t + 1^2} = \sqrt{2}$

$T(t) = \dfrac{r'(t)}{|r'(t)|} = \dfrac{1}{\sqrt{2}} \langle -\sin t, \cos t, 1 \rangle$

$T'(t) = \dfrac{1}{\sqrt{2}} \langle -\cos t, -\sin t, 0 \rangle$

$|T'(t)| = \dfrac{1}{\sqrt{2}} \sqrt{\cos^2 t + \sin^2 t} = \dfrac{1}{\sqrt{2}}$

$N(t) = \dfrac{T'(t)}{|T'(t)|} = \langle -\cos t, -\sin t, 0 \rangle$

$B(t) = T(t) \times N(t)$

$= \dfrac{1}{\sqrt{2}} \begin{vmatrix} i & j & k \\ -\sin t & \cos t & 1 \\ -\cos t & -\sin t & 0 \end{vmatrix}$

$= \dfrac{1}{\sqrt{2}} \langle \sin t, -\cos t, 1 \rangle$

(Stewart, Calculus Early Transcendentals, p. 866)

Velocity and Acceleration

Velocity and Acceleration, Given $r(t)$
Find velocity, speed, and acceleration.
Given the position vector, $r(t)$,
$r(t) = \langle f(t), g(t), h(t) \rangle$

Velocity $r'(t)$	$v(t) = r'(t)$ $v(t) = \langle f', g', h' \rangle$
Speed $\lvert r'(t) \rvert$	$\lvert r'(t) \rvert = \sqrt{(f')^2 + (g')^2 + (h')^2}$
Acceleration $r''(t)$	$a(t) = r''(t)$ $a(t) = \langle f'', g'', h'' \rangle$

Velocity and Position, Given $a(t)$ & Init. Conditions

Find velocity and position vectors. Given:

acceleration $a(t)$ & position vectors $v(0)\,,r(0)$

**Method #1:   Use initial conditions
to solve for constant of integration.**

Velocity $v(t)$	$v(t) \;=\; \int a(t)\,dt$
Position $r(t)$	$r(t) \;=\; \int v(t)\,dt$

**Method #2:   Use initial conditions
as vectors added to definite integral.**

Velocity $v(t)$	$v(t) \;=\; v(t_0) + \int_{t_0}^{t} a(t)\,dt$
Position $r(t)$	$r(t) \;=\; r(t_0) + \int_{t_0}^{t} v(t)\,dt$

Projectile Motion

θ	$=$	Angle of elevation
v_0	$=$	Initial velocity
h_0	$=$	Initial height
g	$=$	Gravity $= 32\frac{ft}{s^2} = 9.8\frac{m}{s^2}$
x	$=$	Horizontal position
x	$=$	$(v_0 \cos\theta)\,t$
y	$=$	Vertical position
y	$=$	$h_0 + (v_0 \sin\theta)\,t - \frac{1}{2}gt^2$

Velocity and Acceleration, Given $r(t)$ -- Ex. 1

Find velocity, acceleration, and speed,

for a particle with position vector:

$$r(t) = \langle\, t^3,\ t^2,\ 1\,\rangle, \quad \text{at } t = 2$$

Velocity $r'(t)$	$r'(t) = \langle\, f',\ g',\ h'\,\rangle$ $r'(t) = \langle\, 3t^2, 2t, 0\,\rangle$
Speed $\lvert r'(t)\rvert$	$\lvert r'(t)\rvert = \sqrt{(f')^2 + (g')^2 + (h')^2}$ $\lvert r'(t)\rvert = \sqrt{9t^4 + 4t^2 + 1}$
Accel. $r''(t)$	$r''(t) = \langle\, f'',\ g'',\ h''\,\rangle$ $r''(t) = \langle\, 6t, 2, 0\,\rangle$
When $t = 2$	Velocity $= r'(2) = \langle\, 12, 4, 0\,\rangle$ Speed $= \lvert r'(2)\rvert = \sqrt{148} = 2\sqrt{37}$ Accel. $= r''(2) = = \langle\, 12, 2, 0\,\rangle$

Velocity and Acceleration, Given $a(t)$ -- Ex. 2a

Moving particle has initial position: $r(0) = \langle 1, 0, 0 \rangle$

with initial velocity: $v(0) = \langle 1, 2, 3 \rangle$

Acceleration is: $a(t) = \langle 4t, 6t, 1 \rangle$

Find its velocity and position at time t.

$v(t) = \int a(t)\, dt = \int \langle 4t, 6t, 1 \rangle\, dt$

$v(t) = \langle 2t^2, 3t^2, t \rangle + C$

$v(0) = \langle 0, 0, 0 \rangle + C$

$\langle 1, 2, 3 \rangle = \langle 0, 0, 0 \rangle + C$

$\langle 1, 2, 3 \rangle = C$

$v(t) = \langle 2t^2, 3t^2, t \rangle + C$

$v(t) = \langle 2t^2, 3t^2, t \rangle + \langle 1, 2, 3 \rangle$

$v(t) = \langle 2t^2 + 1, 3t^2 + 2, t + 3 \rangle$

$r(t) = \int v(t)\, dt$

$r(t) = \int \langle 2t^2 + 1, 3t^2 + 2, t + 3 \rangle\, dt$

Continued ...

Velocity and Acceleration, Given $a(t)$ -- Ex. 2b

Moving particle has initial position: $r(0) = \langle 1, 0, 0 \rangle$

with initial velocity: $v(0) = \langle 1, 2, 3 \rangle$

Acceleration is: $a(t) = \langle 4t, 6t, 1 \rangle$

Find its velocity and position at time t.

Previously Found:

$v(t) = \langle 2t^2 + 1, 3t^2 + 2, t + 3 \rangle$

$r(t) = \int \langle 2t^2 + 1, 3t^2 + 2, t + 3 \rangle \, dt$

$r(t) = \int \langle 2t^2 + 1, 3t^2 + 2, t + 3 \rangle \, dt$

$r(t) = \langle \frac{2}{3}t^3 + t, \ t^3 + 2t, \ \frac{1}{2}t^2 + 3t \rangle + C$

$r(0) = \langle 0, 0, 0 \rangle + C$

$\langle 1, 0, 0 \rangle = \langle 0, 0, 0 \rangle + C$

$\langle 1, 0, 0 \rangle = C$

$r(t) = \langle \frac{2}{3}t^3 + t, \ t^3 + 2t, \ \frac{1}{2}t^2 + 3t \rangle + \langle 1, 0, 0 \rangle$

$r(t) = \langle \frac{2}{3}t^3 + t + 1, \ t^3 + 2t, \ \frac{1}{2}t^2 + 3t \rangle$

Velocity and Acceleration, Given $a(t)$ -- Ex. 3a

Moving particle has initial position: $r(0) = \langle 1, 0, 0 \rangle$

with initial velocity: $v(0) = \langle 1, 2, 3 \rangle$

Acceleration is: $a(t) = \langle 4t, 6t, 1 \rangle$

Find its velocity and position at time t .

Use these equations: (Definite Integrals, no C)

$$v(t) = v(t_0) + \int_{t_0}^{t} a(t)\, dt$$

$$r(t) = r(t_0) + \int_{t_0}^{t} v(t)\, dt$$

$$v(t) = v(t_0) + \int_{t_0}^{t} a(t)\, dt$$

$$v(t) = \langle 1, 2, 3 \rangle + \int_{0}^{t} \langle 4t, 6t, 1 \rangle\, dt$$

$$v(t) = \langle 1, 2, 3 \rangle + \langle 2t^2, 3t^2, t \rangle$$

$$v(t) = \langle 2t^2 + 1, 3t^2 + 2, t + 3 \rangle$$

Continued ...

Velocity and Acceleration, Given $a(t)$ -- Ex. 3b

Moving particle has initial position: $r(0) = \langle 1, 0, 0 \rangle$

with initial velocity: $v(0) = \langle 1, 2, 3 \rangle$

Acceleration is: $a(t) = \langle 4t, 6t, 1 \rangle$

Find its velocity and position at time t.

Use these equations: (Definite Integrals, no C)

$$v(t) = v(t_0) + \int_{t_0}^{t} a(t)\, dt$$

$$r(t) = r(t_0) + \int_{t_0}^{t} v(t)\, dt$$

$$r(t) = r(t_0) + \int_{t_0}^{t} v(t)\, dt$$

$$r(t) = \langle 1, 0, 0 \rangle + \int_{0}^{t} \langle 2t^2 + 1, 3t^2 + 2, t + 3 \rangle\, dt$$

$$r(t) = \langle 1, 0, 0 \rangle + \langle \tfrac{2}{3}t^3 + t, t^3 + 2t, \tfrac{1}{2}t^2 + 3t \rangle$$

$$r(t) = \langle \tfrac{2}{3}t^3 + t + 1, t^3 + 2t, \tfrac{1}{2}t^2 + 3t \rangle$$

Projectile Motion -- Ex. 4a

A projectile is fired at a speed of $200 \, \frac{m}{s}$

At an elevation of 30° from a position of 50 m above

the ground. Where does the projectile hit the ground

and with what speed?

Find the time when the projectile hits the ground.
That's when $y = 0$.

$y = h_0 + (v_0 \sin \theta) t - \frac{1}{2} g t^2$

$y = 50 + (200)(\sin 30°) t - \frac{1}{2} (9.8) t^2$

$y = 50 + (200) \left(\frac{1}{2} \right) t - \frac{1}{2} (9.8) t^2$

$y = 50 + 100 t - 4.9 t^2$

$0 = -4.9 t^2 + 100 t + 50$

$t = \dfrac{-100 \pm \sqrt{100^2 - 4(-4.9)(50)}}{2(-4.9)}$

$t = -0.5, \ 20.9$ 　　　　　Disregard negative time.

Projectile Motion -- Ex. 4b

A projectile is fired at a speed of $200 \frac{m}{s}$

At an elevation of $30°$ from a position of $50\ m$ above the ground. Where does the projectile hit the ground and with what speed?

Previously Found:

Projectile hits the ground when $t = 20.9\ s$

Find the horizontal position when the projectile hits ground (when $t = 20.9\ s$)

---

x = Horizontal position

$x = (v_0 \cos \theta)\, t$

$x = (200)\left(\frac{\sqrt{3}}{2}\right) t \quad \approx (173.2)\, t$

$x = (173.2)(20.9) \quad \approx 3620\ m$

Continued ...

Projectile Motion -- Ex. 4c
A projectile is fired at a speed of $200\ \frac{m}{s}$ At an elevation of 30° from a position of 50 m above   the ground. Where does the projectile hit the ground   and with what speed?

Previously, found projectile hits ground when: $\quad t = 20.9\ s,\qquad x = 3620\ m,\qquad y = 0$
To find the speed, we need the velocity vector.   Use: $v(t) = r'(t) \qquad$ at $t = 20.9\ s$
$r(t) = \langle\,(173.2\,t\,),\ (\,50 + 100t - 4.9\,t^2\,)\,\rangle$
$v(t)\ =\ r'(t)\ =\ \langle 173.2\,,\ 100 - 9.8\,t\,\rangle$ $v(\,20.9\,)\ =\ \langle 173.2\,,\ 100 - 9.8\,(20.9)\,\rangle$ $v(\,20.9\,)\ =\ \langle 173.2\,,\ -104.8\ \rangle$ Speed $=\ \mid v(t)\mid\ =\ \sqrt{173.2^2 + (-104.8)^2}$ Speed $=\ 202.4\ \frac{m}{s}$

Partial Derivatives

Functions of Several Variables

Functions of Two Variables
$f(x, y) = $ Function of two variables.
$z = f(x, y)$ $x, y = $ Independent Variables $z = $ Dependent Variable

Domain	The set of (x, y) pairs for which the function is defined.
Range	The set of numbers on a real line (e.g. The z axis).

Functions of Two Variables -- Level Curves

Level Curves of a function of 2 variables are the curves

with equations: $f(x, y) = k$

$k =$ a constant in the range of f

For example, contour maps.

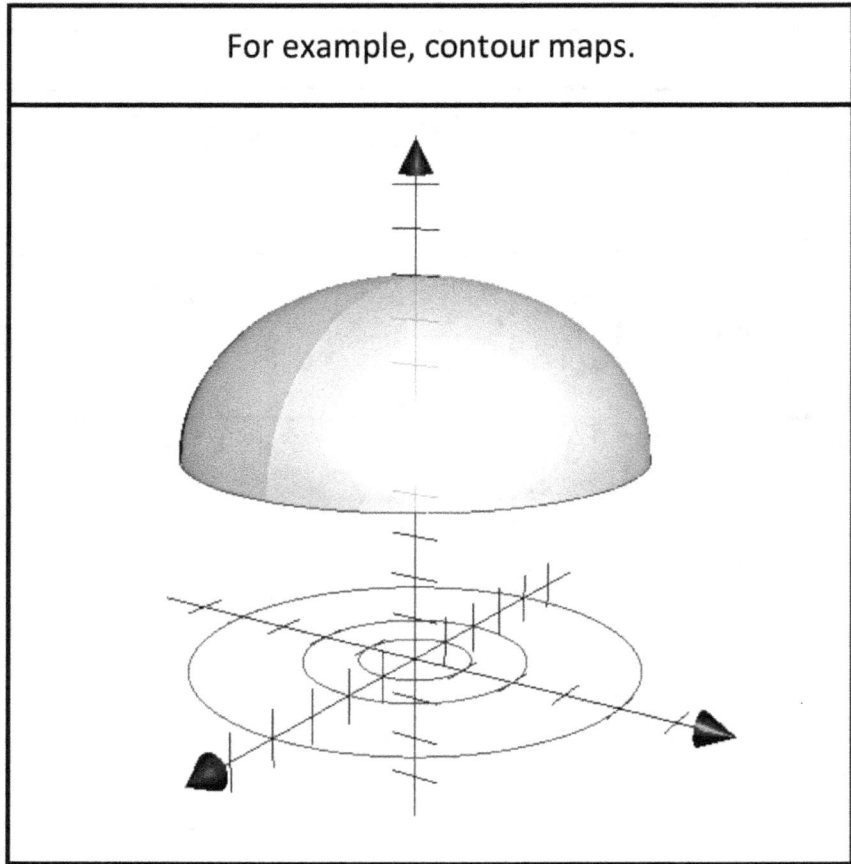

Functions of Two Variables -- Ex. 1

For the given function, evaluate $f(3,4)$ then sketch the domain.

$$f(x,y) = \frac{\sqrt{x+y-3}}{x-2}$$

$$f(3,4) = \frac{\sqrt{3+4-3}}{3-2} = \frac{\sqrt{3}}{1} = \sqrt{3}$$

Domain restrictions:

- Can't divide by zero, so $x \neq 2$.
- Can't take square root of a negative number, so $x + y - 3 \geq 0$

$x \neq 2$

$y \geq 3 - x$

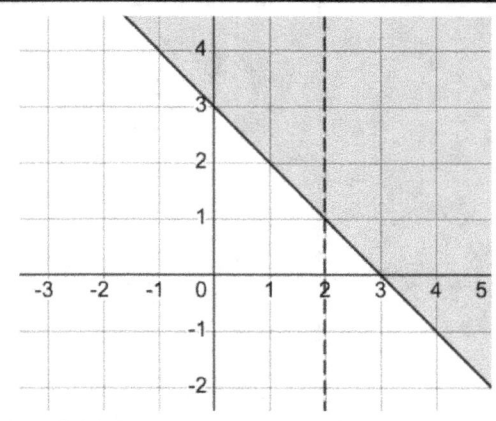

$$D = \{\,(x,y)\mid x+y-3 \geq 0,\ x \neq 2\,\}$$

Functions of Two Variables -- Ex. 2

For the given function, evaluate $f(3,2)$

then sketch the domain.

$$f(x,y) = 2x \ln(y^2 - x)$$

$f(3,2) = 2(3) \ln(4-3) = 0$

Domain restrictions:

- $y^2 - x > 0 \quad \rightarrow \quad x < y^2$

$D = Domain = \{ (x,y) \mid x < y^2 \}$

Domain

$x < y^2$

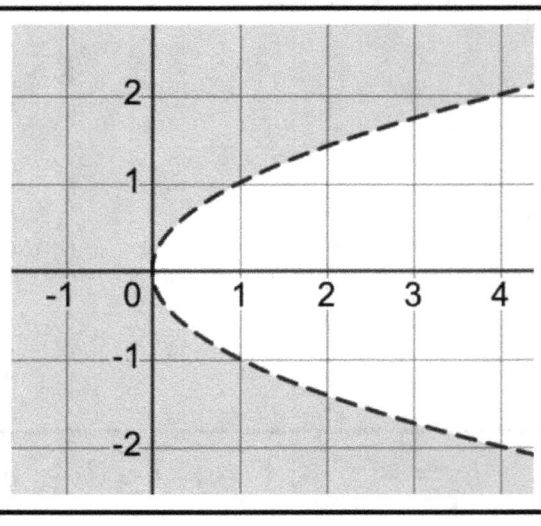

Functions of Two Variables -- Ex. 3

Find the domain and range for

$$f(x,y) = \sqrt{16 - x^2 - y^2}$$

Domain	$16 - x^2 - y^2 \geq 0$ $x^2 + y^2 \leq 16$ $D = \{ (x,y) \mid x^2 + y^2 \leq 16 \}$ 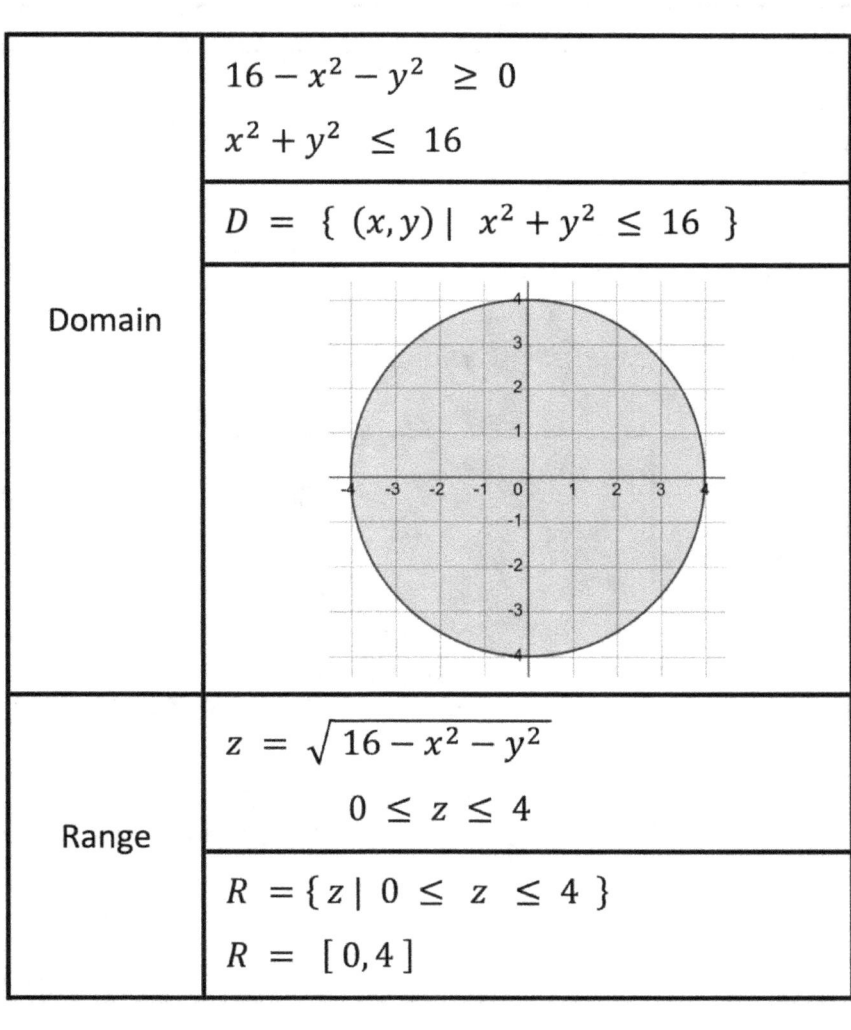
Range	$z = \sqrt{16 - x^2 - y^2}$ $0 \leq z \leq 4$ $R = \{z \mid 0 \leq z \leq 4 \}$ $R = [0,4]$

Functions of Two Variables -- Ex. 4

Sketch the graph of

$$g(x, y) = \sqrt{16 - x^2 - y^2}$$

$$z = \sqrt{16 - x^2 - y^2} \qquad \text{With } z \geq 0$$

$$z^2 = 16 - x^2 - y^2$$

$$x^2 + y^2 + z^2 = 16$$

Top half of sphere with radius $= 4$

Sketch	

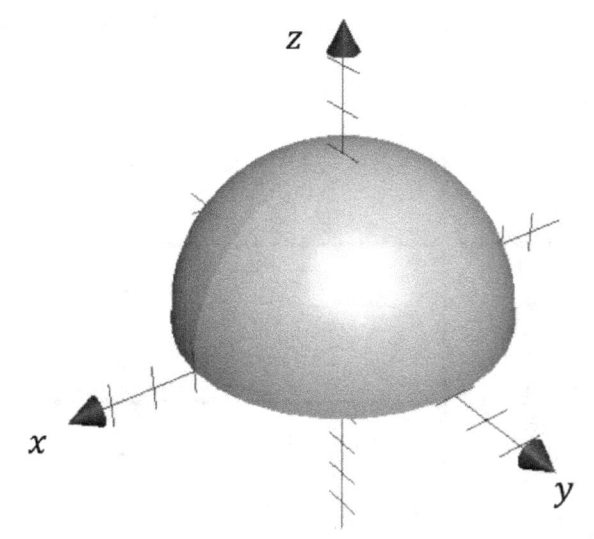

Functions of Two Variables -- Ex. 5

Sketch the graph of

$$g(x, y) = \sin x + \sin y$$

Equation of graph: $z = \sin x + \sin y$

Use a 3D graphing utility. (e.g. Grapher)

| Sketch | |

Functions of Two Variables -- Level Curves -- Ex. 6

Sketch the level curves for the function

$$f(x,y) = 5 - x - 2y \qquad \text{For: } k = -5, 0, 5, 10$$

Set $f(x,y) = k$	$5 - x - 2y = k$
Rearrange	$0 = x + 2y + (k - 5)$

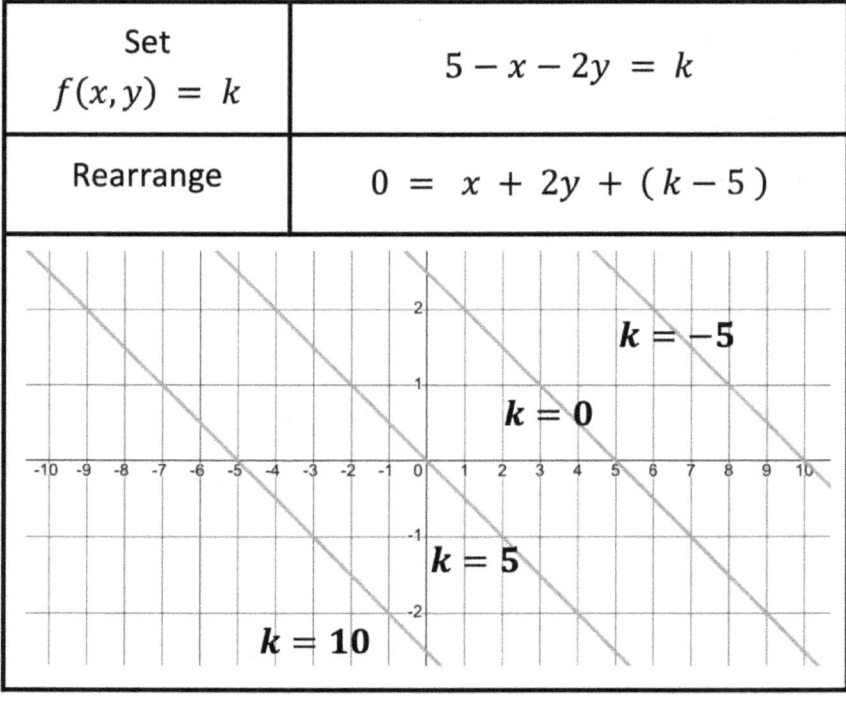

Functions of Two Variables -- Level Curves -- Ex. 7

Sketch the level curves for the function

$$f(x,y) = \sqrt{16 - x^2 - y^2} \qquad \text{For: } k = 0, 2, 3, 4$$

Set $f(x,y) = k$	$\sqrt{16 - x^2 - y^2} = k$
Rearrange	$0 = x^2 + y^2 + (k^2 - 16)$

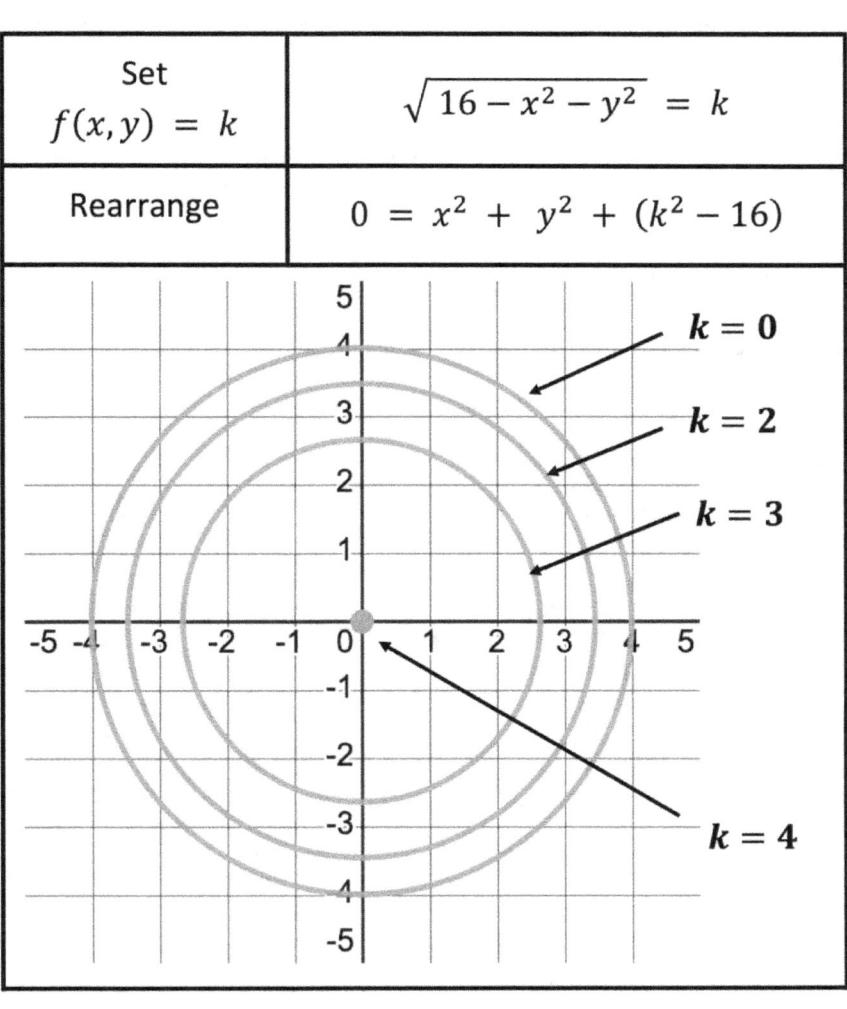

Functions of Two Variables -- Level Curves -- Ex. 8a

Sketch some level curves for the function

$$f(x,y) = 2x^2 - y^2 - 1$$

Set $f(x,y) = k$	$2x^2 - y^2 - 1 = k$
Traces are Hyperbolas	$2x^2 - y^2 = (k+1)$ $\dfrac{x^2}{\frac{1}{2}(k+1)} - \dfrac{y^2}{(k+1)} = 1$

$k = 1$

$k = 3$

$k = 7$

Functions of Two Variables -- Level Curves -- Ex. 8b

Sketch some level curves for the function

$$f(x, y) = 2x^2 - y^2 - 1$$

Extra ...	
Previously found: Some level curves for $k = 1, 3, 7$	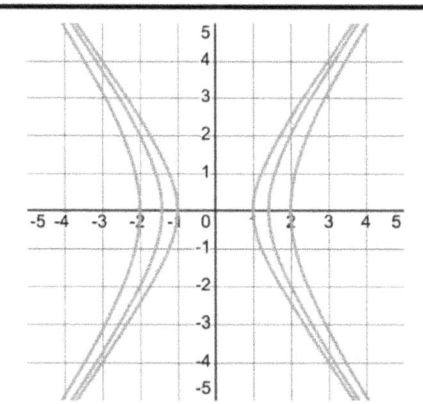
Use a graphing utility to sketch of the surface. $$z = 2x^2 - y^2 - 1$$	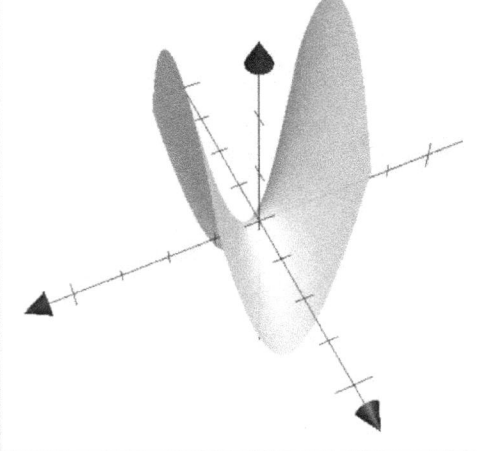

Limits and Continuity

Limits of $f(x, y)$
$$\lim_{(x,y) \to (a,b)} f(x,y) = L$$
The limit of $f(x, y)$, as (x, y) approaches (a, b), is L.

For the limit to exist, the same limit must be obtained when approaching (a, b) from any direction, or path.

If

$$\lim_{(x,y) \to (a,b)} f(x,y) = L_1 \quad \text{along path } C_1$$

And

$$\lim_{(x,y) \to (a,b)} f(x,y) = L_2 \quad \text{along path } C_2$$

Then,

$$\lim_{(x,y) \to (a,b)} f(x,y) \quad \text{DNE}$$

DNE = Does Not Exist!

Continuity of $f(x, y)$
If $\quad$ $f(x, y)$ is continuous at (a, b) $$\lim_{(x,y) \to (a,b)} f(x,y) = f(a,b)$$ The function is continuous   at every point in its' domain, D.

Polynomial and Rational functions   are continuous on their domains.
If 2 functions, f and g, are continuous then, their   composites, fog and gof, are also continuous.

Limits of $f(x,y)$ -- Ex. 1

Does $\displaystyle\lim_{(x,y)\to(0,0)} f(x,y)$ exist
If $\quad f(x,y) = \dfrac{xy}{x^2+y^2}$?

Check path where $y = 0$	$\displaystyle\lim_{(x,0)\to(0,0)} \dfrac{x(0)}{x^2+0^2}$
	$\displaystyle\lim_{(x,0)\to(0,0)} \dfrac{0}{x^2} = 0$
Check path where $x = 0$	$\displaystyle\lim_{(0,y)\to(0,0)} \dfrac{(0)y}{0^2+y^2}$
	$\displaystyle\lim_{(0,y)\to(0,0)} \dfrac{0}{y^2} = 0$
Check path where $x = y$	$\displaystyle\lim_{(y,y)\to(0,0)} \dfrac{(y)y}{y^2+y^2}$
	$\displaystyle\lim_{(y,y)\to(0,0)} \dfrac{y^2}{2y^2} = \dfrac{1}{2}$
$L_1 \neq L_2$	Therefore, the limit DNE

Limits of $f(x, y)$ -- Ex. 2

Does	$\lim\limits_{(x,y) \to (0,0)} f(x, y)$ exist
If	$f(x, y) = \dfrac{xy^2}{x^2 + y^2}$?

Check path where $y = mx$	$\lim\limits_{(x,mx) \to (0,0)} \dfrac{x\,(mx)^2}{x^2 + (mx)^4}$
	$\lim\limits_{(x,mx) \to (0,0)} \dfrac{m^2 x^3}{x^2 + m^4 x^4}$
	$\lim\limits_{(x,mx) \to (0,0)} \dfrac{m^2 x}{1 + m^4 x^2} = 0$
Check path where $x = y^2$	$\lim\limits_{(y^2,y) \to (0,0)} \dfrac{(y^2)\,y^2}{y^4 + y^4}$
	$\lim\limits_{(y^2,y) \to (0,0)} \dfrac{y^4}{2y^4} = \dfrac{1}{2}$
$L_1 \neq L_2$	Therefore, the limit DNE

Limits of $f(x,y)$ -- Ex. 3		

Where is the function continuous?

$$h(x,y) = \tan^{-1}\left(\frac{y}{x}\right)$$

Rewrite as composite of two continuous functions	$f(x,y) = \frac{y}{x}$ $g(t) = \tan^{-1}(t)$ $h(x,y) = g(f(x,y))$ $h(x,y) = g \circ f$
Domains	$D_f = \{(x,y)\mid x \neq 0\}$ $D_g = \{t \mid t \in \mathbb{R}\}$ $D_h = \{(x,y)\mid x \neq 0\}$
Where is $h(x,y)$ continuous?	$h(x,y)$ is continuous everywhere in its domain, D_h.

Stewart, Calculus Early Transcendentals, p. 909)

Partial Derivatives

Partial Derivatives of $f(x, y)$
$$f_x(x, y) = \lim_{h \to 0} \frac{f(x+h, \ y) - f(x, \ y)}{h}$$ $$f_y(x, y) = \lim_{h \to 0} \frac{f(x, \ y+h) - f(x, \ y)}{h}$$
$$f_x = \frac{\partial f}{\partial x} = \frac{\partial}{\partial x} f(x, y) = \frac{\partial z}{\partial x} = D_x f$$ $$f_y = \frac{\partial f}{\partial y} = \frac{\partial}{\partial y} f(x, y) = \frac{\partial z}{\partial y} = D_y f$$

To find f_x , regard y as a constant and differentiate $f(x, y)$ WRT x
To find f_y , regard x as a constant and differentiate $f(x, y)$ WRT y

WRT = With Respect To

First & Second Partial Derivatives of $z = f(x, y)$

First Partial Derivatives

$$f_x = \frac{\partial f}{\partial x} = \frac{\partial}{\partial x} f(x, y) = \frac{\partial z}{\partial x} = D_x f$$

$$f_y = \frac{\partial f}{\partial y} = \frac{\partial}{\partial y} f(x, y) = \frac{\partial z}{\partial y} = D_y f$$

Second Partial Derivatives

$$(f_x)_x = f_{xx} = \frac{\partial}{\partial x}\left(\frac{\partial f}{\partial x}\right) = \frac{\partial^2 f}{\partial x^2}$$

$$(f_x)_y = f_{xy} = \frac{\partial}{\partial y}\left(\frac{\partial f}{\partial x}\right) = \frac{\partial^2 f}{\partial y\,\partial x}$$

$$(f_y)_x = f_{yx} = \frac{\partial}{\partial x}\left(\frac{\partial f}{\partial y}\right) = \frac{\partial^2 f}{\partial x^2}$$

$$(f_y)_y = f_{yy} = \frac{\partial}{\partial y}\left(\frac{\partial f}{\partial y}\right) = \frac{\partial^2 f}{\partial y^2}$$

Partial Derivatives of $f(x,y)$ -- Ex. 1

Find: $f_x(1,1)$ and $f_y(1,1)$

Given: $f(x,y) = 9 - 3x^2 - y^2$

$f_x(1,1)$	$f_x(x,y) = -6x$ $f_x(1,1) = -6(1) = -6$
$f_y(1,1)$	$f_y(x,y) = -2y$ $f_y(1,1) = -2(1) = -2$
Extra	$f(x,y) = z = 9 - 3x^2 - y^2$ $f(1,1) = z = 9 - 3(1)^2 - (1)^2$ $z = 9 - 3 - 1 = 5$
	At the point $(x,y,z) = (1,1,5)$ Slope of tangent in x-direction $= -6$ Slope of tangent in y-direction $= -2$

Partial Derivatives of $f(x, y)$ -- **Ex. 2**

Find: $f_x(1,2)$ and $f_y(1,2)$

Given: $f(x,y) = 9 - 3x^2 - y^2$

$f_x(1,2)$	$f_x(x,y) = -6x$ $f_x(1,2) = -6(1) = -6$
$f_y(1,2)$	$f_y(x,y) = -2y$ $f_y(1,2) = -2(2) = -4$
Extra	$f(x,y) = z = 9 - 3x^2 - y^2$ $f(1,2) = z = 9 - 3(1)^2 - (2)^2$ $z = 9 - 3 - 4 = 2$
	At the point $(x, y, z) = (1, 2, 2)$ Slope of tangent in x-direction $= -6$ Slope of tangent in y-direction $= -4$

Partial Derivatives of $f(x,y)$ -- Ex. 3	
Find: $\dfrac{\partial f}{\partial x}$ and $\dfrac{\partial f}{\partial y}$	
Given: $f(x,y) = \sin\left(\dfrac{x}{1+y}\right)$	

$\dfrac{\partial f}{\partial x}$	$\dfrac{\partial f}{\partial x} = \cos\left(\dfrac{x}{1+y}\right) \cdot \dfrac{\partial f}{\partial x}\left(\dfrac{x}{1+y}\right)$ $\dfrac{\partial f}{\partial x} = \cos\left(\dfrac{x}{1+y}\right) \cdot \left(\dfrac{1}{1+y}\right)$ Chain Rule
$\dfrac{\partial f}{\partial y}$	$\dfrac{\partial f}{\partial y} = \cos\left(\dfrac{x}{1+y}\right) \cdot \dfrac{\partial f}{\partial y}\left(\dfrac{x}{1+y}\right)$ $\dfrac{\partial f}{\partial y} = \cos\left(\dfrac{x}{1+y}\right) \cdot \dfrac{\partial f}{\partial y}\left(x(1+y)^{-1}\right)$ $\dfrac{\partial f}{\partial y} = -\cos\left(\dfrac{x}{1+y}\right) \cdot \left(x(1+y)^{-2}\right)$ $\dfrac{\partial f}{\partial y} = -\cos\left(\dfrac{x}{1+y}\right) \cdot \dfrac{x}{(1+y)^2}$

(Stewart, Calculus Early Transcendentals, p. 917)

Partial Derivatives of $f(x,y)$ -- Ex. 4

Find: $\dfrac{\partial z}{\partial x}$ and $\dfrac{\partial z}{\partial y}$

Given: $x^3 + y^3 + z^3 + 6xyz = 1$

Use implicit differentiation. Note that $z = f(x,y)$.

$\dfrac{\partial z}{\partial x}$	$\dfrac{\partial}{\partial x}\left[x^3 + y^3 + z^3 + 6xyz\right] = \dfrac{\partial}{\partial x}[1]$ $3x^2 + 3z^2\dfrac{\partial z}{\partial x} + \left[6xy\dfrac{\partial z}{\partial x} + 6yz\right] = 0$ $\dfrac{\partial z}{\partial x} = \dfrac{-3x^2 - 6yz}{3z^2 + 6xy} = -\dfrac{x^2 + 2yz}{z^2 + 2xy}$
$\dfrac{\partial z}{\partial y}$	$\dfrac{\partial}{\partial y}\left[x^3 + y^3 + z^3 + 6xyz\right] = \dfrac{\partial}{\partial y}[1]$ $3y^2 + 3z^2\dfrac{\partial z}{\partial y} + \left[6xy\dfrac{\partial z}{\partial y} + 6xz\right] = 0$ $\dfrac{\partial z}{\partial y} = \dfrac{-3y^2 - 6xz}{3z^2 + 6xy} = -\dfrac{y^2 + 2xz}{z^2 + 2xy}$

(Stewart, Calculus Early Transcendentals, p. 917)

Second Partial Derivatives of $f(x, y)$ -- Ex. 5

Find the second partial derivatives for

$$f(x,y) = x^5 + x^4y^3 - 2y^2$$

$$f_x = \frac{\partial f}{\partial x} = 5x^4 + 4x^3y^3$$

$$f_y = \frac{\partial f}{\partial y} = 3x^4y^2 - 4y$$

$$(f_x)_x = f_{xx} = 20x^3 + 12x^2y^3$$

$$(f_x)_y = f_{xy} = 12x^3y^2$$

$$(f_y)_x = f_{yx} = 12x^3y^2$$

$$(f_y)_y = f_{yy} = 6x^4y - 4$$

Tangent Planes and Linear Approx.

Tangent Plane of a 3D Surface at a Point	
Surface	$z = f(x, y)$
Point on Surface	$P_0(x_0, y_0, z_0)$
Tangent Plane	$\Delta z \;=\; f_x \Delta x \,+\, f_y \Delta y$ $z - z_0 \;=\; f_x(x_0, y_0)(x - x_0)$ $\qquad\qquad + \; f_y(x_0, y_0)(y - y_0)$
Notes	Δ = Delta = Change The Greek symbol, Δ, represents the change in a variable.

Tangent Plane of a 3D Surface at a Point -- Ex. 1

Find the equation of the tangent plane to the given surface & point on the surface.

Surface: $z = 2x^2 + y^2$ Point: $(1, 1, 3)$

$f_x = 4x$ $f_x(1,1) = 4$	$f_y = 2y$ $f_y(1,1) = 2$

$$\Delta z = f_x \Delta x + f_y \Delta y$$
$$z - 3 = 4(x - 1) + 2(y - 1)$$
$$z - 3 = 4x - 4 + 2y - 2$$
$$z = 4x + 2y - 3$$

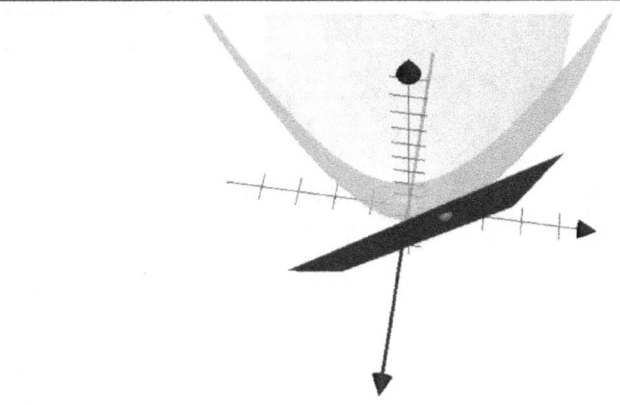

(Stewart, Calculus Early Transcendentals, p. 928)

Tangent Plane of a 3D Surface at a Point -- Ex. 2

Find the equation of the tangent plane to the given surface & point on the surface.

Surface: $z = 2x^2 + y^2$ Point: $(1, 2, 3)$

$f_x = 4x$	$f_y = 2y$
$f_x(1, 2) = 4(1) = 4$	$f_y(1, 2) = 2(2) = 4$

$$\Delta z = f_x \Delta x + f_y \Delta y$$
$$z - 3 = 4(x - 1) + 4(y - 2)$$
$$z - 3 = 4x - 4 + 4y - 8$$
$$z = 4x + 2y - 9$$

Linear Approximation of f at (a, b)
Tangent Plane Approximation -- Ex. 3

Given: $f(x, y) = xe^{2y}$ is differentiable at $(1, 0)$
Find its linearization there.
Then, use it to approximate $f(1.1, 0.1)$

$f_x = e^{2y}$	$f_y = 2xe^{2y}$
$f_x(1, 0) = e^{2 \cdot 0} = 1$	$f_y(1, 0) = 2e^0 = 2$

$L(x, y) = f(1,0) + f_x(1,0) \cdot \Delta x + f_y(1,0) \cdot \Delta y$

$L(x, y) = (1)e^{2(0)} + (1)(x - 1) + 2(y - 0)$

$L(x, y) = 1 + x - 1 + 2y$

$L(x, y) = x + 2y$

$L(1.1, 0.1) = 1.1 + 2(0.1) = 1.3$

Compare approximation to actual ...

$f(x, y) = xe^{2y}$

$f(1.1, 0.1) = (1.1)e^{2(0.1)} = 1.34$

Total Differential
For: $z = f(x, y)$

dz = Total Differential

$dz = f_x(x, y)\, dx + f_y(x, y)\, dy$

$dz = \dfrac{\delta z}{\delta x}\, dx + \dfrac{\delta z}{\delta y}\, dy$

Total Differential -- Example

A right circular cone is measured as having the following radius & height.

$$r = 8 \ cm \quad \text{and} \quad h = 20 \ cm$$

The possible error is $0.1 \ cm$ for each measurement.

Estimate the maximum error
in the calculated volume of the cone.

$$V = \frac{1}{3}\pi r^2 h = f(r, h)$$

$$dV = \frac{\partial V}{\partial r} dr + \frac{\partial V}{\partial h} dh$$

$$dV = \left(\frac{2}{3}\pi rh\right) dr + \left(\frac{1}{3}\pi r^2\right) dh$$

$$dV \approx (335.1)(.1) + (67)(.1)$$

$$dV \approx (33.5) + (6.7) \approx 40 \ cm^3$$

The Chain Rule

The Chain Rule

Recall, if $y = f(x)$ and $x = f(t)$

Then: $\dfrac{dy}{dt} = \dfrac{dy}{dx}\dfrac{dx}{dt}$

Chain Rule	
$z = f(x,y)$ $x = g(t)$ $y = h(t)$	$\dfrac{dz}{dt} = \dfrac{\partial z}{\partial x}\dfrac{dx}{dt} + \dfrac{\partial z}{\partial y}\dfrac{dy}{dt}$
Chain Rule $z = f(x,y)$ $x = g(s,t)$ $y = h(s,t)$	$\dfrac{\partial z}{\partial s} = \dfrac{\partial z}{\partial x}\dfrac{\partial x}{\partial s} + \dfrac{\partial z}{\partial y}\dfrac{\partial y}{\partial s}$ $\dfrac{\partial z}{\partial t} = \dfrac{\partial z}{\partial x}\dfrac{\partial x}{\partial t} + \dfrac{\partial z}{\partial y}\dfrac{\partial y}{\partial t}$

The Chain Rule -- General

Suppose u is a function of n variables

And each of the variables is a function of m variables.

$$u = u(x_1, x_2, x_3 \ldots x_n)$$

Each $\quad x_j = x(t_1, t_2, t_3 \ldots t_m)$

$$\frac{\partial u}{\partial t_i} = \frac{\partial u}{\partial x_1}\frac{\partial x_1}{\partial t_i} + \frac{\partial u}{\partial x_2}\frac{\partial x_2}{\partial t_i} + \cdots + \frac{\partial u}{\partial x_n}\frac{\partial x_n}{\partial t_i}$$

For each $i = 1, 2, \ldots m$

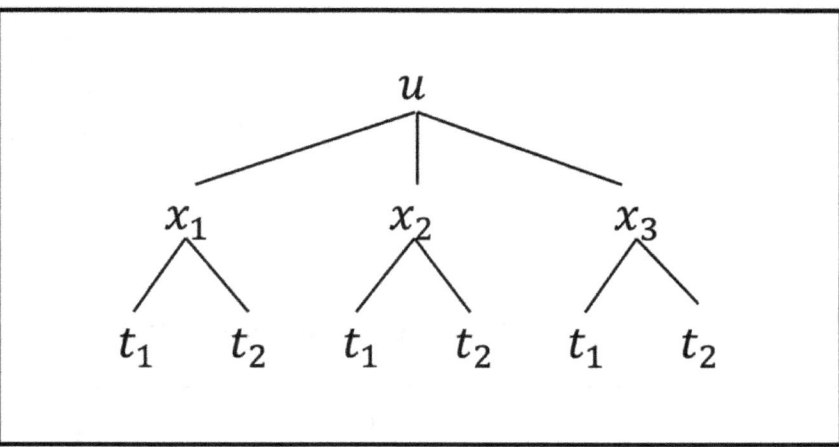

The Chain Rule -- Implicit Differentiation

If: $\qquad F(x,y) = F(x, f(x)) = 0$

Then: $\qquad F_x\, x' + F_y\, y_x = 0$

$$F_x\, (1) + F_y\, y_x = 0$$

$$y_x = -\frac{F_x}{F_y}$$

$F(x,y)$ must be defined & continuous in the domain.

If: $\qquad F(x,y,z) = 0$

Then: $\qquad z_x = -\frac{F_x}{F_z}$

$$z_y = -\frac{F_y}{F_z}$$

$F(x,y,z)$ must be defined & continuous in domain.

The Chain Rule -- Ex. 1a

Given: $u = xy + y^2z^3$

With: $x = rs^2e^t$

$$y = rs^3e^{-t}$$

$$z = r^3s \sin t$$

Find: $\dfrac{\partial u}{\partial s}$ when $r = 1$, $s = 2$, $t = 0$

When:
$r = 1$
$s = 2$
$t = 0$

$x = rs^2e^t$	$= 4$
$y = rs^3e^{-t}$	$= 8$
$z = r^3s \sin t$	$= 0$

$$\frac{\partial u}{\partial s} = \frac{\partial u}{\partial x}\frac{\partial x}{\partial s} + \frac{\partial u}{\partial y}\frac{\partial y}{\partial s} + \frac{\partial u}{\partial z}\frac{\partial z}{\partial s}$$

$$\frac{\partial u}{\partial s} = (y)(2rse^t)$$

$$+ (x + 2yz^3)(3rs^2e^{-t})$$

$$+ (3y^2z^2)(r^3 \sin t)$$

The Chain Rule -- Ex. 1b

Given: $u = xy + y^2z^3$

With: $x = rs^2e^t$

$$y = rs^3e^{-t}$$

$$z = r^3s\sin t$$

Find: $\dfrac{\partial u}{\partial s}$ when $r = 1,\ s = 2,\ t = 0$

Previously Found	$(x, y, z) = (4, 8, 0)$
	When: $(r, s, t) = (1, 2, 0)$

$$\frac{\partial u}{\partial s} = \frac{\partial u}{\partial x}\frac{\partial x}{\partial s} + \frac{\partial u}{\partial y}\frac{\partial y}{\partial s} + \frac{\partial u}{\partial z}\frac{\partial z}{\partial s}$$

$$\frac{\partial u}{\partial s} = (y)(2rse^t)$$

$$+ (x + 2yz^3)(3rs^2e^{-t})$$

$$+ (3y^2z^2)(r^3\sin t)$$

$$\frac{\partial u}{\partial s} = (8)(4) + (4)(12) + (0)(0)$$

$$= 32 + 48 + 0 = 80$$

The Chain Rule -- Ex. 2a

Given: $z = f(x, y)$

With: $x = r^2 + s^2$ and $y = 2rs$

Find: $\dfrac{\partial z}{\partial r}$ and $\dfrac{\partial^2 z}{\partial r^2}$

$$\frac{\partial z}{\partial r} = \frac{\partial z}{\partial x}\frac{\partial x}{\partial r} + \frac{\partial z}{\partial y}\frac{\partial y}{\partial r}$$

$$\frac{\partial z}{\partial r} = \frac{\partial z}{\partial x}(2r) + \frac{\partial z}{\partial y}(2s)$$

$$\frac{\partial}{\partial r}\left[\frac{\partial z}{\partial r}\right] = \frac{\partial}{\partial r}\left[\frac{\partial z}{\partial x}(2r) + \frac{\partial z}{\partial y}(2s)\right]$$

$$\frac{\partial^2 z}{\partial r^2} = \left[\frac{\partial}{\partial r}\left(\frac{\partial z}{\partial x}\right)(2r) + \left(\frac{\partial z}{\partial x}\right)(2)\right] +$$

$$+ \left[\frac{\partial}{\partial r}\left(\frac{\partial z}{\partial y}\right)(2s) + \left(\frac{\partial z}{\partial y}\right)(0)\right]$$

$$\frac{\partial^2 z}{\partial r^2} = 2r\frac{\partial}{\partial r}\left(\frac{\partial z}{\partial x}\right) + 2\left(\frac{\partial z}{\partial x}\right) + 2s\frac{\partial}{\partial r}\left(\frac{\partial z}{\partial y}\right)$$

(Stewart, Calculus Early Transcendentals, p. 941)

The Chain Rule -- Ex. 2b

Given: $z = f(x, y)$

With: $x = r^2 + s^2$ and $y = 2rs$

Find: $\dfrac{\partial z}{\partial r}$ and $\dfrac{\partial^2 z}{\partial r^2}$

$$\frac{\partial^2 z}{\partial r^2} = 2r\frac{\partial}{\partial r}\left(\frac{\partial z}{\partial x}\right) + 2\left(\frac{\partial z}{\partial x}\right) + 2s\frac{\partial}{\partial r}\left(\frac{\partial z}{\partial y}\right)$$

$\dfrac{\partial}{\partial r}\left(\dfrac{\partial z}{\partial x}\right)$	$= \dfrac{\partial}{\partial x}\left(\dfrac{\partial z}{\partial x}\right)\dfrac{\partial x}{\partial r} + \dfrac{\partial}{\partial y}\left(\dfrac{\partial z}{\partial x}\right)\dfrac{\partial y}{\partial r}$
	$= \dfrac{\partial^2 z}{\partial x^2} \cdot \dfrac{\partial x}{\partial r} + \dfrac{\partial^2 z}{\partial y \partial x} \cdot \dfrac{\partial y}{\partial r}$
	$= \dfrac{\partial^2 z}{\partial x^2}(2r) + \dfrac{\partial^2 z}{\partial y \partial x}(2s)$
$\dfrac{\partial}{\partial r}\left(\dfrac{\partial z}{\partial y}\right)$	$= \dfrac{\partial}{\partial x}\left(\dfrac{\partial z}{\partial y}\right)\dfrac{\partial x}{\partial r} + \dfrac{\partial}{\partial y}\left(\dfrac{\partial z}{\partial y}\right)\dfrac{\partial y}{\partial r}$
	$= \dfrac{\partial^2 z}{\partial x \partial y} \cdot \dfrac{\partial x}{\partial r} + \dfrac{\partial^2 z}{\partial y^2} \cdot \dfrac{\partial y}{\partial r}$
	$= \dfrac{\partial^2 z}{\partial x \partial y}(2r) + \dfrac{\partial^2 z}{\partial y^2}(2s)$

The Chain Rule -- Ex. 2c

Given: $z = f(x, y)$

With: $x = r^2 + s^2$ and $y = 2rs$

Find: $\dfrac{\partial z}{\partial r}$ and $\dfrac{\partial^2 z}{\partial r^2}$

$\dfrac{\partial}{\partial r}\left(\dfrac{\partial z}{\partial x}\right)$	$= \dfrac{\partial^2 z}{\partial x^2}(2r) + \dfrac{\partial^2 z}{\partial y\,\partial x}(2s)$
$\dfrac{\partial}{\partial r}\left(\dfrac{\partial z}{\partial y}\right)$	$= \dfrac{\partial^2 z}{\partial x\,\partial y}(2r) + \dfrac{\partial^2 z}{\partial y^2}(2s)$

$$\frac{\partial^2 z}{\partial r^2} = 2\left(\frac{\partial z}{\partial x}\right) + 2r\frac{\partial}{\partial r}\left(\frac{\partial z}{\partial x}\right) + 2s\frac{\partial}{\partial r}\left(\frac{\partial z}{\partial y}\right)$$

$$\frac{\partial^2 z}{\partial r^2} = 2\left(\frac{\partial z}{\partial x}\right) + 2r\left[\frac{\partial^2 z}{\partial x^2}(2r) + \frac{\partial^2 z}{\partial y\,\partial x}(2s)\right]$$

$$+ 2s\left[\frac{\partial^2 z}{\partial x\,\partial y}(2r) + \frac{\partial^2 z}{\partial y^2}(2s)\right]$$

$$\frac{\partial^2 z}{\partial r^2} = 2\left(\frac{\partial z}{\partial x}\right) + 4r^2\frac{\partial^2 z}{\partial x^2} + 8rs\frac{\partial^2 z}{\partial y\,\partial x} + 4s^2\frac{\partial^2 z}{\partial y^2}$$

The Chain Rule -- Ex. 3a
(Example 1, using different notation)

Given: $z = f(x, y)$

With: $x = r^2 + s^2$ and $y = 2rs$

Find: $\dfrac{\partial z}{\partial r}$ and $\dfrac{\partial^2 z}{\partial r^2}$

Find: z_r and z_{rr}

$z_r = z_x\, x_r + z_y\, y_r$

$z_r = z_x\, (2r) + z_y\, (2s)$

$\dfrac{\partial}{\partial r}[\, z_r\,] = \dfrac{\partial}{\partial r}\big[\, z_x\,(2r) + z_y\,(2s)\,\big]$

$z_{rr} = [\, z_{xr}(2r) + z_x(2)\,] +$

$\qquad\qquad + \big[\, z_{yr}(2s) + z_y(0)\,\big]$

$z_{rr} = 2r\, z_{xr} + 2\, z_x + 2s\, z_{yr}$

The Chain Rule -- Ex. 3b
(Example 1, using different notation)

Given: $z = f(x, y)$

With: $x = r^2 + s^2$ and $y = 2rs$

Find: $\dfrac{\partial z}{\partial r}$ and $\dfrac{\partial^2 z}{\partial r^2}$

Find: z_r and z_{rr}

Previously Found	$z_{rr} = 2r\, z_{xr} + 2\, z_x + 2s\, z_{yr}$
z_{xr}	$z_{xr} = z_{xx} \cdot x_r + z_{xy} \cdot y_r$ $z_{xr} = z_{xx}\,(2r) + z_{xy}\,(2s)$
z_{yr}	$z_{yr} = z_{xy} \cdot x_r + z_{yy} \cdot y_r$ $z_{yr} = z_{xy}\,(2r) + z_{yy}\,(2s)$

Continued …

The Chain Rule -- Ex. 3c
(Example 1, using different notation)

Given: $z = f(x, y)$

With: $x = r^2 + s^2$ and $y = 2rs$

Find: $\dfrac{\partial z}{\partial r}$ and $\dfrac{\partial^2 z}{\partial r^2}$

Find: z_r and z_{rr}

z_{rr}	$z_{rr} = 2r\, z_{xr} + 2\, z_x + 2s\, z_{yr}$
z_{xr}	$z_{xr} = z_{xx}\,(2r) + z_{xy}\,(2s)$
z_{yr}	$z_{yr} = z_{xy}\,(2r) + z_{yy}\,(2s)$

$$z_{rr} = 2\, z_x + 2r\, z_{xr} + 2s\, z_{yr}$$

$$z_{rr} = 2\, z_x + 2r\left[z_{xx}\,(2r) + z_{xy}\,(2s) \right]$$

$$+ 2s\left[z_{xy}\,(2r) + z_{yy}\,(2s) \right]$$

$$z_{rr} = 2\, z_x + 4r^2\, z_{xx} + 8rs\, z_{xy} + 4s^2\, z_{yy}$$

The Chain Rule -- Implicit Differentiation -- Ex. 4

Given: $x^2 + y^3 + 6xyz = 1$

Find: $\dfrac{\partial z}{\partial x}$ and $\dfrac{\partial z}{\partial y}$ $\qquad (z_x \ \& \ z_y)$

$F(x, y, z) \ = \ x^2 + y^3 + 6xyz - 1$

$F(x, y, z) \ = \ 0 \qquad\qquad$ (Required Format)

Get partial derivatives for $F(x, y, z)$	$F_x \ = \ 2x + 6yz$
	$F_y \ = \ 3y^2 + 6xz$
	$F_z \ = \ 6xy$

$$z_x \ = \ -\frac{F_x}{F_z} \ = \ -\frac{2x + 6yz}{6xy} \ = \ -\frac{x + 3yz}{3xy}$$

$$z_y \ = \ -\frac{F_y}{F_z} \ = \ -\frac{3y^2 + 6xz}{6xy} \ = \ -\frac{y^2 + 2xz}{2xy}$$

Directional Derivatives & Gradient Vector

Directional Derivative

$D_u f(x, y)$ = Directional Derivative
In the direction of u

$D_u f(x, y) = f_x a + f_y b$

$D_u f(x, y) = \langle f_x, f_y \rangle \cdot \langle a, b \rangle$

Where:

$f(x, y)$ = Differentiable function

$u = \langle a, b \rangle$ = Any unit vector

The Gradient Vector

$$D_u f(x, y) = \nabla f(x, y) \cdot u$$

$$D_u f(x, y) = \langle f_x, f_y \rangle \cdot u$$

$$D_u f(x, y) = \nabla f(x, y) \cdot u$$

$$D_u f(x, y) = \langle f_x, f_y \rangle \cdot u$$

Directional Derivative -- Maximum

The directional derivative is maximized when

the unit vector, u ,has the same direction

as the gradient vector, $\nabla f(x)$.

The maximum value

of the directional derivative, $D_u f(x)$,

Is: $|\nabla f(x)|$

$$D_u f \;=\; \nabla f \cdot u$$

$$D_u f \;=\; |\nabla f|\,|u|\cos\theta$$

$$D_u f \;=\; |\nabla f|\cos\theta$$

Maximum occurs when $\theta = 0$

$\theta =$ The angle between the unit vector

and the gradient vector.

$$\cos(0) = 1$$

The Maximum $D_u f \;=\; |\nabla f|$

Directional Derivative -- Ex. 1

Find the directional derivative of function

$$f(x, y, z) = x \sin(yz)$$

At point $(1, 3, 0)$ in the direction of $v = \langle 1, 2, 3 \rangle$

$$\nabla f(x, y, z) = \langle f_x, f_y, f_z \rangle$$

$$\nabla f(x, y, z) = \langle \sin(yz), xz \cos(yz), xy \cos(yz) \rangle$$

$$|v| = \sqrt{1^2 + 2^2 + 3^2} = \sqrt{14}$$

Unit Vector $= u = \dfrac{1}{\sqrt{14}} \langle 1, 2, 3 \rangle$

$$D_u f(a, b, c) = \nabla f(a, b, c) \cdot u$$

$$D_u f(1, 3, 0) = \nabla f(1, 3, 0) \cdot \frac{1}{\sqrt{14}} \langle 1, 2, 3 \rangle$$

$$= \langle \sin(0), \ 0 \cos(0), \ 3 \cos(0) \rangle \cdot \frac{1}{\sqrt{14}} \langle 1, 2, 3 \rangle$$

$$= \langle 0, 0, 3 \rangle \cdot \frac{1}{\sqrt{14}} \langle 1, 2, 3 \rangle$$

$$= \frac{1}{\sqrt{14}} (0 + 0 + 9) = \frac{9}{\sqrt{14}}$$

Directional Derivative -- Maximum -- Ex. 2a

For: $f(x, y) = xe^y$

Find the rate of change of f at point $P(2, 0)$

In the direction from P to $Q\left(\frac{1}{2}, 2\right)$

Also: In what direction does f have the greatest rate of change? What is this maximum rate of change?

$\nabla f(x, y) = \langle f_x, f_y \rangle = \langle e^y, x \rangle$

$\nabla f(2, 0) = \langle e^0, 2 \rangle = \langle 1, 2 \rangle$

$\overrightarrow{PQ} = \langle \frac{1}{2} - 2,\ 2 - 0 \rangle = \langle -\frac{3}{2},\ 2 \rangle$

$|\overrightarrow{PQ}| = \sqrt{\frac{9}{4} + 4} = \sqrt{\frac{25}{4}} = \frac{5}{2}$

$u = \frac{\overrightarrow{PQ}}{|\overrightarrow{PQ}|} = \frac{2}{5}\langle -\frac{3}{2},\ 2 \rangle = \langle -\frac{3}{5},\ \frac{4}{5} \rangle$

Continued ...

Directional Derivative -- Maximum -- Ex. 2b

For: $f(x,y) = xe^y$

Find the rate of change of f at point $P(2,0)$

In the direction from P to $Q\left(\frac{1}{2}, 2\right)$

Also: In what direction does f have the greatest rate

of change?   What is this maximum rate of change?

Previously Found	$\nabla f(2,0) = \langle 1,2 \rangle$ $u = \langle -\frac{3}{5}, \frac{4}{5} \rangle$

$D_u f = \nabla f \cdot u$

$D_u f(2,0) = \nabla f(2,0) \cdot u$

$D_u f(2,0) = \langle 1,2 \rangle \cdot \langle -\frac{3}{5}, \frac{4}{5} \rangle$

$D_u f(2,0) = -\frac{3}{5} + \frac{8}{5} = \frac{5}{5} = 1$

Continued ...

Directional Derivative -- Maximum -- Ex. 2c

For: $f(x, y) = xe^y$

Find the rate of change of f at point $P(2, 0)$

In the direction from P to $Q\left(\frac{1}{2}, 2\right)$

Also: In what direction does f have the greatest rate of change? What is this maximum rate of change?

Previously Found	$\nabla f(2, 0) = \langle 1, 2 \rangle$ $u = \langle -\frac{3}{5}, \frac{4}{5} \rangle$ $D_u f(2, 0) = 1$
f increases fastest in direction of gradient vector	$\nabla f(2, 0) = \langle 1, 2 \rangle$
Max. Rate of Change	$\lvert \nabla f(2, 0) \rvert = \lvert \langle 1, 2 \rangle \rvert = \sqrt{5}$

(Stewart, Calculus Early Transcendentals, p. 952)

<u>Maximum and Minimum Values</u>

Max & Min Values for $f(x,y)$ -- Critical Point

If

f has a local max or min at (a,b)

and f_x and f_y exist there

Then

(a,b) = **Critical Point**

Where: $f_x = 0$ and $f_y = 0$

Local Max At (a,b)	$f(x,y) \leq F(a,b)$ Near (a,b)
Local Min At (a,b)	$f(x,y) \geq F(a,b)$ Near (a,b)

Max & Min Values for $f(x, y)$ **-- 2nd Derivative Test**

If

$$f_x(a, b) = f_y(a, b) = 0$$

Then

$$(a, b) = \underline{\textbf{Critical Point}}$$

$$D = D(a, b) \quad \text{Important calculation}$$

$$D = f_{xx}(a, b) \cdot f_{yy}(a, b) - \big[f_{xy}(a, b) \big]^2$$

$$D = \begin{vmatrix} f_{xx} & f_{xy} \\ f_{yy} & f_{yx} \end{vmatrix} = f_{xx} f_{yy} - (f_{xy})^2$$

D	f_{xx}	$f(a, b)$
Positive	Positive	Local Min
	Negative	Local Max
Negative	---	Saddle Pt.

ABSOLUTE Max & Min Values for $f(x, y)$

If

f is continuous on a **closed** ,
bounded domain, D, in $\mathbb{R}^2$,

Then

f has an absolute max & min
value, in the domain, D.

Closed Sets	
Not Closed	

How to find Absolute Max & Min values:

- Find values of f at critical pts. in D

- Find extreme values of f on closed boundary of D.

- Compare above values to find absolute max & min.

Max & Min Values for $f(x, y)$ -- Ex. 1a

Find the extreme values of the function.

$$f(x, y) = x^2 + y^2 - 2x - 4y + 16$$

Find critical point (CP)	$f_x = 0$ $2x - 2 = 0$ $x = 1$	$f_y = 0$ $2y - 4 = 0$ $y = 2$
Evaluate function at CP To find extrema	$z = f(x, y)$ $z = f(1, 2)$ $z = 1^2 + 2^2 - 2(1) - 4(2) + 16$ $z = 11$	

Complete the square for more information:

$$z = x^2 + y^2 - 2x - 4y + 16$$
$$z - 16 = x^2 - 2x + y^2 - 4y$$
$$z - 11 = (x^2 - 2x + 1) + (y^2 - 4y + 4)$$
$$z = (x - 1)^2 + (y - 2)^2 + 11$$

For all values of x and y the function:

$$z = f(x, y) \geq 11 \quad \rightarrow \quad f(1, 2) = 11 \quad \text{is a local MIN}$$

Max & Min Values for $f(x, y)$ **-- Ex. 1b**

Find the extreme values of the function.

$$f(x, y) = x^2 + y^2 - 2x - 4y + 16$$

Previously Found	$f(1, 2) = 11$ is a local MIN Point: $(x, y, z) = (1, 2, 11)$
	Completed the Square to find: $f(x, y) = (x - 1)^2 + (y - 2)^2 + 11$ Elliptic Paraboloid

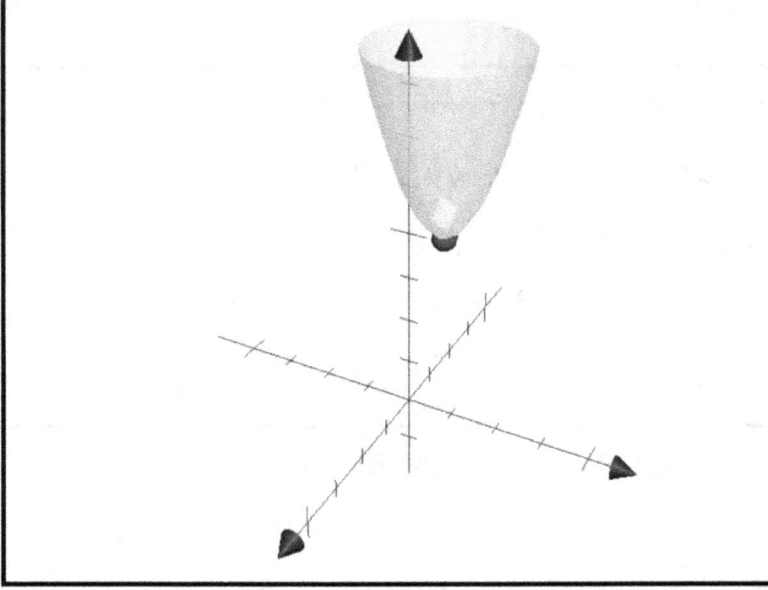

Max & Min Values for $f(x, y)$ -- Ex. 2a

Find the extreme values of the function.

$$f(x, y) = x^2 - y^2$$

Find critical point (CP)	$f_x = 0$ $2x = 0$ $x = 0$	$f_y = 0$ $-2y = 0$ $y = 0$
Evaluate function at CP To find extrema	$z = f(x, y)$ $z = f(0, 0)$ $z = 0^2 - 0^2 = 0$	

Critical point when: $\quad z = f(0, 0) = 0$

Or: $\qquad\qquad\qquad (x, y, z) = (0, 0, 0)$

$f(x, 0) = x^2 - 0 \geq 0$

$f(0, y) = 0 - y^2 \leq 0$

So, near point $(0, 0, 0)$ some values are positive and some values are negative.

Therefore: $f(0, 0) = 0$ is **NOT** an extrema.

Max & Min Values for $f(x,y)$ -- Ex. 2b

Find the extreme values of the function.

$$f(x,y) = x^2 - y^2$$

Previously Found	Critical point at: $(x,\ y,\ z) = (0,0,0)$
	Near point (0,0,0) some values are positive and some values are negative. Therefore: $f(0,0) = 0$ is **NOT** an extrema.

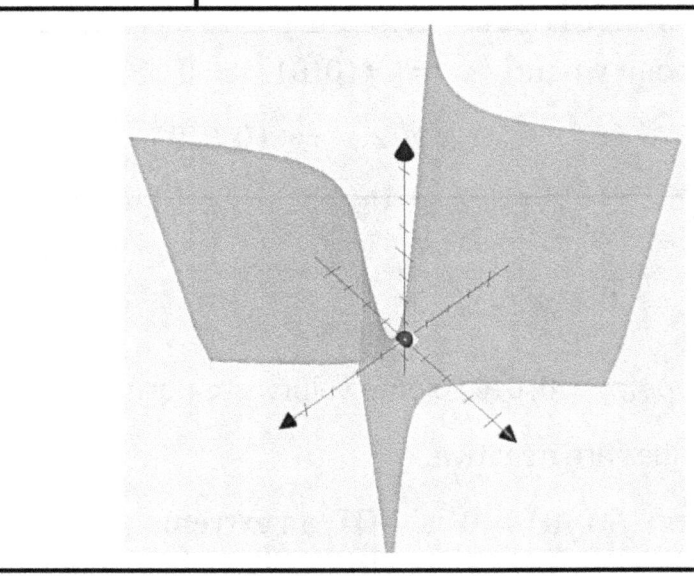

Max & Min Values -- 2nd Derivative Test -- Ex. 3a

Find the local maximum and minimum values and saddle points of

$$f(x,y) = x^4 + y^4 - 4xy + 1$$

Find critical points (CPs)	$f_x = 0$ $4x^3 - 4y = 0$ $x^3 - y = 0$	$f_y = 0$ $4y^3 - 4x = 0$ $y^3 - x = 0$

$x = 0 \quad \rightarrow \quad y = 0 \quad \rightarrow$ Point $(0,0)$

$x = 1 \quad \rightarrow \quad y = 1 \quad \rightarrow$ Point $(1,1)$

$x = -1 \quad \rightarrow \quad y = -1 \quad \rightarrow$ Point $(-1,-1)$

Three CPs: $(a,b) = (0,0), (1,1), (-1,-1)$

$f_{xx} = 12x^2$	$f_{yy} = 12y^2$	$f_{xy} = -4$

$$D = f_{xx}\, f_{yy} - \left[\, f_{xy}\, \right]^2$$

Calculate the value of *"D"* for all critical points.

(Stewart, Calculus Early Transcendentals, p. 961)

Max & Min Values -- 2nd Derivative Test -- Ex. 3b

Find the local maximum and minimum values and saddle points of
$$f(x, y) = x^4 + y^4 - 4xy + 1$$

Previously Found	Three Critical Points: $(a, b) = (0,0), (1,1), (-1,-1)$ $f_{xx} = 12x^2, \ f_{yy} = 12y^2, \ f_{xy} = -4$
Get ready to calculate D for all CPs	$D = f_{xx} f_{yy} - [f_{xy}]^2$ $D = (12x^2)(12y^2) - [-4]^2$ $D = 144x^2 y^2 - 16$

(a, b)	D	f_{xx}	$f(a, b)$
$(0, 0)$	-16	0	1
$(1, 1)$	128	12	-1
$(-1, -1)$	128	12	-1

Max & Min Values -- 2nd Derivative Test -- Ex. 3c

Find the local maximum and minimum values and saddle points of

$$f(x,y) = x^4 + y^4 - 4xy + 1$$

(a,b)	D	f_{xx}	$f(a,b)$
$(0,0)$	-16	0	1
$(1,1)$	128	12	-1
$(-1,-1)$	128	12	-1

(a,b)	Conclusions
$(0,0)$	$D < 0 \quad \rightarrow \quad$ Saddle Point At $(0,0,1)$
$(1,1)$	$f_{xx} > 0 \quad \rightarrow \quad$ Local MIN At $(1,1,-1)$
$(-1,-1)$	$f_{xx} > 0 \quad \rightarrow \quad$ Local MIN At $(-1,-1,-1)$

Max & Min Values -- 2nd Derivative Test -- Ex. 3d

Find the local maximum and minimum values and saddle points of
$$f(x,y) = x^4 + y^4 - 4xy + 1$$

(a, b)	Conclusions		
$(0,0)$	$D < 0$	$\rightarrow$	Saddle Point At $(0,0,1)$
$(1,1)$	$f_{xx} > 0$	$\rightarrow$	Local MIN At $(1,1,-1)$
$(-1,-1)$	$f_{xx} > 0$	$\rightarrow$	Local MIN At $(-1,-1,-1)$

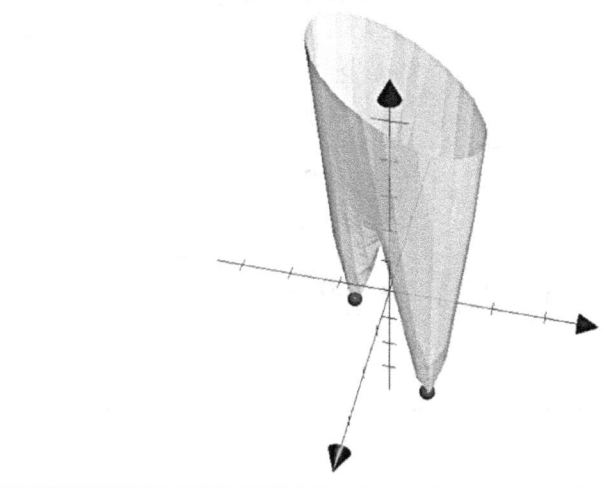

Max & Min Values -- 2nd Derivative Test -- Ex. 4a

A rectangular box without a lid, is to be made
From $12 \, m^2$ of cardboard.
Find the maximum volume of such a box.

Let length, width, and height be x, y, z	
Use constraint to get $z = f(x, y)$	$Area = 12$ $2xz + 2yz + xy = 12$ $z(2x + 2y) = 12 - xy$ $z = \dfrac{12 - xy}{2x + 2y}$
Goal	Goal is to maximize volume so express volume in terms of just x & y

(Stewart, Calculus Early Transcendentals, p. 964)

Max & Min Values -- 2nd Derivative Test -- Ex. 4b

A rectangular box without a lid, is to be made
From $12 \ m^2$ of cardboard.
Find the maximum volume of such a box.

Previously found $z = f(x, y)$	$z \ = \ \dfrac{12 \ - \ xy}{2x \ + \ 2y}$

$$V = xyz = xy \left[\frac{12 - xy}{2x + 2y} \right] = \frac{12xy - x^2y^2}{2(x + y)}$$

$$V_x = \frac{2(x+y)\left[12y - 2xy^2\right] - [\,2\,](12xy - x^2y^2)}{4(x + y)^2}$$

$$V_x = \frac{(x+y)\left[12y - 2xy^2\right] - (12xy - x^2y^2)}{2(x + y)^2}$$

$$V_x = \frac{12xy - 2x^2y^2 + 12y^2 - 2xy^3 - (12xy - x^2y^2)}{2(x + y)^2}$$

$$V_x = \frac{-x^2y^2 + 12y^2 - 2xy^3}{2(x + y)^2} = \frac{y^2(12 - 2xy - x^2)}{2(x + y)^2}$$

$$V_y = \frac{x^2(12 - 2xy - y^2)}{2(x + y)^2} \qquad \text{(Work not shown)}$$

Max & Min Values -- 2nd Derivative Test -- Ex. 4c

A rectangular box without a lid, is to be made
From $12 \ m^2$ of cardboard.
Find the maximum volume of such a box.

Previously found $z = f(x, y)$	$z \ = \ \dfrac{12 \ - \ xy}{2x \ + \ 2y}$
	$V_x \ = \ \dfrac{y^2(12 - 2xy - x^2)}{2(x + y)^2}$
	$V_y \ = \ \dfrac{x^2(12 - 2xy - y^2)}{2(x + y)^2}$

Extrema occurs when: $\quad V_x \ = \ V_y \ = \ 0$

$V_x \ = \ \dfrac{y^2(12 - 2xy - x^2)}{2(x+y)^2} \ = \ 0$

$y^2(12 \ - \ 2xy \ - \ x^2) \ = \ 0$

$x^2 + 2xy - 12 \ = \ 0$ $\qquad \boxed{\begin{array}{c} y > 0 \\ y \neq 0 \end{array}}$

$(x + 6)(x - 2) \ = \ 0$

$x \ = \ -6, \ 2 \quad \rightarrow \quad x = 2$ $\qquad \boxed{x > 0}$

Max & Min Values -- 2nd Derivative Test -- Ex. 4d

A rectangular box without a lid, is to be made
From $12\ m^2$ of cardboard.
Find the maximum volume of such a box.

Previously found
$z = f(x, y)$

$$z = \frac{12 - xy}{2x + 2y}$$

$$V_x = \frac{y^2(12 - 2xy - x^2)}{2(x + y)^2}$$

$$V_y = \frac{x^2(12 - 2xy - y^2)}{2(x + y)^2}$$

Extrema when $x = 2$

$$V_y = \frac{x^2(12 - 2xy - y^2)}{2(x+y)^2} = 0$$

$$y^2 + 2xy - 12 = 0$$

$$(y + 6)(y - 2) = 0$$

$$y = 2$$

$$\boxed{\begin{array}{l} x, y > 0 \\ x, y \neq 0 \end{array}}$$

$$z = \frac{12 - xy}{2x + 2y} = \frac{12 - (2)(2)}{2(2) + 2(2)} = \frac{8}{8} = 1$$

Max & Min Values -- 2nd Derivative Test -- Ex. 4e

A rectangular box without a lid, is to be made
From $12\ m^2$ of cardboard.
Find the maximum volume of such a box.

Previously Found	Extrema when $x = 2, \quad y = 2, \quad z = 1$

We could use the 2nd derivative test to show that this extrema is a local max.

From the physical nature of this problem, we can simply argue that there must be an absolute maximum volume, which occurs at the critical point.

$$(x, y, z) \ = \ (2, 2, 1)$$

So, the maximum volume is:

$$V \ = \ xyz \ = \ (2)(2)(1) \ = \ 4\ m^3$$

(Stewart, Calculus Early Transcendentals, p. 965)

ABSOLUTE Max & Min Values -- Ex. 5a

Find the absolute Max and Min values

For: $f(x,y) = x^2 - 2xy + 2y$

on the domain, specified as a rectangle, D.

$$D = \{(x,y) \mid 0 \le x \le 3, \ 0 \le y \le 2\}$$

Since f is a polynomial, it is continuous
on the closed and bounded domain, D.

$f_x = 2x - 2y$	$f_y = -2x + 2$
$0 = 2x - 2y$	$0 = -2x + 2$
$x = y$	$x = 1$

The only critical point at $(x,y) = (1,1)$

$z = x^2 - 2xy + 2y$

$z = 1^2 - 2(1)(1) + 2(1) = 1$

(Stewart, Calculus Early Transcendentals, p. 966)

ABSOLUTE Max & Min Values -- Ex. 5b

Find the absolute Max and Min values

For: $f(x, y) = x^2 - 2xy + 2y$

on the domain, specified as a rectangle, D.

$$D = \{ (x, y) \mid 0 \le x \le 3, \ 0 \le y \le 2 \}$$

Previously Found	Critical Point at $(x, y, z) = (1, 1, 1)$
Sketch Domain, D	

For the sketch: $(0, 2)$ L_3 $(3, 2)$ L_4 L_2 $(0, 0)$ L_1 $(3, 0)$

L_1	$y = 0$	$f(x, 0) = x^2$
L_2	$x = 3$	$f(3, y) = 9 - 4y$
L_3	$y = 2$	$f(x, 2) = x^2 - 4x + 4$
L_4	$x = 0$	$f(0, y) = 2y$

ABSOLUTE Max & Min Values -- Ex. 5c

Find the absolute Max and Min values

For: $f(x,y) = x^2 - 2xy + 2y$

on the domain, specified as a rectangle, D.

$$D = \{ (x,y) \mid 0 \leq x \leq 3, \ 0 \leq y \leq 2 \}$$

Find max & min on boundary of domain.
Remember to **evaluate within boundary**.

Line	Equation	Min ; Max
L_1	$f(x,0) = x^2$	0 ; 9
L_2	$f(3,y) = 9 - 4y$	1 ; 9
L_3	$f(x,2) = x^2 - 4x + 4$	0 ; 4
L_4	$f(0,y) = 2y$	0 ; 4

Compare CP value(s) to the boundary min and max values to get absolute min and max.

Abs. MIN: 0   at $(x,y) = (0,0),(1,2),(2,2)$

Abs. MAX: 9   at $(x,y) = (3,0)$

Lagrange Multipliers

Lagrange Multipliers
Used to find extreme values of a function: $f(x,y)$ subject to constraint: $g(x,y) = k$

How to find Lagrange Multipliers	Find 3 unknowns (x, y, λ) By solving these 3 equations: $$f_x = \lambda g_x$$ $$f_y = \lambda g_y$$ $$g(x,y) = k$$
How to use Lagrange Multipliers	Evaluate the function, f, at all (x, y) pairs found above. Largest is the Max. Smallest is the Min.
Based on these equations.	$$\nabla f(x,y,z) = \lambda \nabla g(x,y,z)$$ $$g(x,y,z) = k$$

Lagrange Multipliers -- Ex. 1a

A rectangular box without a lid, is to be made from $12\ m^2$ of cardboard.

Find the maximum volume of such a box.

Note	This problem was done in previous section, without Lagrange Multipliers.
Let length, width, and height be $x,\ y,\ z$	
Write constraint In the form $g(x,y,z)\ =k$	$Area\ =\ 12$ $2xz\ +\ 2yz\ +\ xy\ =\ 12$ $g(x,y,z)\ =\ k$
Goal is to maximize Volume	$V\ =\ xyz$

(Stewart, Calculus Early Transcendentals, p. 973)

Lagrange Multipliers -- Ex. 1b

A rectangular box without a lid, is to be made

from $12\ m^2$ of cardboard.

Find the maximum volume of such a box.

Previously Found	$V = f(x,y,z)\ =\ xyz$ $g(x,y,z)\ =\ k$ $2xz\ +\ 2yz\ +\ xy\ =\ 12$

$V_x = \lambda g_x$	$yz\ =\ \lambda(2z+y)$
$V_y = \lambda g_y$	$xz\ =\ \lambda(2z+x)$
$V_z = \lambda g_z$	$xy\ =\ \lambda(2x+2y)$
Solve for λ	$\lambda\ =\ \dfrac{yz}{2z+y}\ =\ \dfrac{xz}{2z+x}\ =\ \dfrac{xy}{2x+2y}$

$$\frac{yz}{2z+y}\ =\ \frac{xz}{2z+x}$$

$$y(2z+x)\ =\ x(2z+y)\quad \rightarrow \quad y = x$$

Lagrange Multipliers -- Ex. 1c

A rectangular box without a lid, is to be made

from $12 \ m^2$ of cardboard.

Find the maximum volume of such a box.

Previously Found	$\lambda = \dfrac{yz}{2z + y} = \dfrac{xz}{2z + x} = \dfrac{xy}{2x + 2y}$ $x = y$

$\dfrac{xz}{2z + x} = \dfrac{xy}{2x + 2y}$

$y(2z + x) = z(2x + 2y)$

$2yz + xy - 2xz - 2yz = 0$

$xy - 2xz = x(y - 2z) = 0$

$y = 2z \qquad\qquad \rightarrow \qquad x = 2z$

$2xz + 2yz + xy = 12$

$2(2z)z + 2(2z)z + (2z)(2z) = 12$

$4z^2 + 4z^2 + 4z^2 = 12$

$12z^2 = 12 \qquad\qquad \rightarrow \qquad z = 1$

Lagrange Multipliers -- Ex. 1d

A rectangular box without a lid, is to be made

from $12\ m^2$ of cardboard.

Find the maximum volume of such a box.

Previously Found	$\lambda = \dfrac{yz}{2z+y} = \dfrac{xz}{2z+x} = \dfrac{xy}{2x+2y}$ $x = y = 2$ $z = 1$
Extrema when	$(x,\ y,\ z) = (2,2,1)$
Maximum volume of box.	$V = xyz$ $V = (2)(2)(1) = 4m^3$ This agrees with the answer, previously found.

(Stewart, Calculus Early Transcendentals, p.974)

Lagrange Multipliers -- Ex. 2a

Find extreme values

Of the function: $f(x, y) = 2x + y^2$

On the circle: $x^2 + y^2 = 5$

Constraint	$g(x, y) = x^2 + y^2 = 5$
Function	$f(x, y) = 2x + y^2$

$f_x = \lambda g_x$ $2 = \lambda 2x$ $1 = \lambda x \rightarrow \lambda, x \neq 0$	$f_y = \lambda g_y$ $2y = \lambda 2y$ $\lambda = 1$ or $y = 0$
$\lambda = 1 \rightarrow x = 1$	$(1)^2 + y^2 = 5$ $y = \pm 2$
$y = 0 \rightarrow$	$x = \pm\sqrt{5}$

Evaluate function at critical points.	$f(1, \pm 2) = 6$ $f(\sqrt{5}, 0) = 2\sqrt{5}$ $f(-\sqrt{5}, 0) = -2\sqrt{5}$

Lagrange Multipliers -- Ex. 2b

Find extreme values

Of the function: $f(x, y) = 2x + y^2$

On the circle: $x^2 + y^2 = 5$

Constraint	$g(x,y) = x^2 + y^2 = 5$
Function	$f(x,y) = 2x + y^2$
Extreme Values: Function evaluated at CPs.	$z = f(1, \pm 2) = 6$ $z = f(\sqrt{5}, 0) = 2\sqrt{5}$ $z = f(-\sqrt{5}, 0) = -2\sqrt{5}$

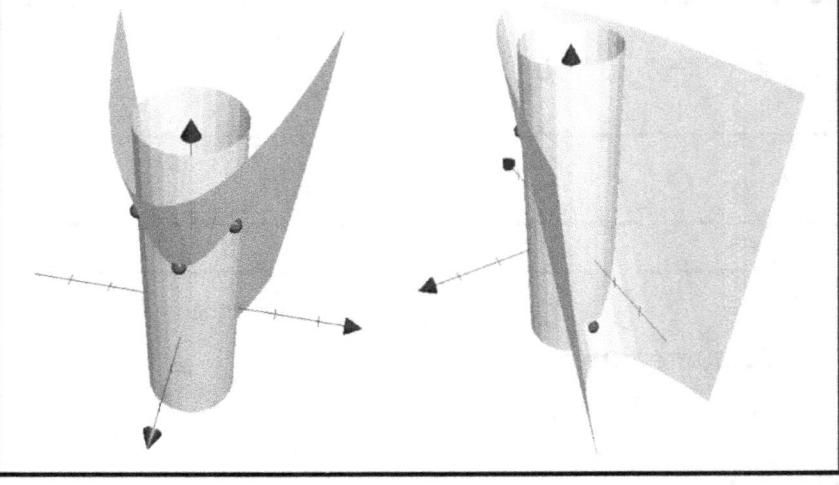

Lagrange Multipliers -- Ex. 3a

Find extreme values

Of the function: $f(x,y) = x^2 + 2y^2$

On the circle: $x^2 + y^2 = 1$

Constraint	$g(x,y) = x^2 + y^2 = 1$
Function	$f(x,y) = x^2 + 2y^2$

$f_x = \lambda g_x$	$f_y = \lambda g_y$
$2x = \lambda 2x$	$4y = \lambda 2y$
$\lambda = 1$ or $x = 0$	$\lambda = 2$ or $y = 0$
$x = 0$	$0^2 + y^2 = 1 \rightarrow y = \pm 1$
$y = 0$	$x^2 + 0^2 = 1 \rightarrow x = \pm 1$

Evaluate function at CPs.	$z = f(0, \pm 1) = 2$ MAX
	$z = f(\pm 1, 0) = 1$ MIN

(Stewart, Calculus Early Transcendentals, p. 974)

Lagrange Multipliers -- Ex. 3b

Find extreme values

Of the function: $f(x, y) = x^2 + 2y^2$

On the circle: $x^2 + y^2 = 1$

Constraint	$g(x, y) = x^2 + y^2 = 1$
Function	$f(x, y) = x^2 + 2y^2$
Evaluate function at critical points.	$z = f(0, \pm 1) = 2$ $z = f(\pm 1, 0) = 1$

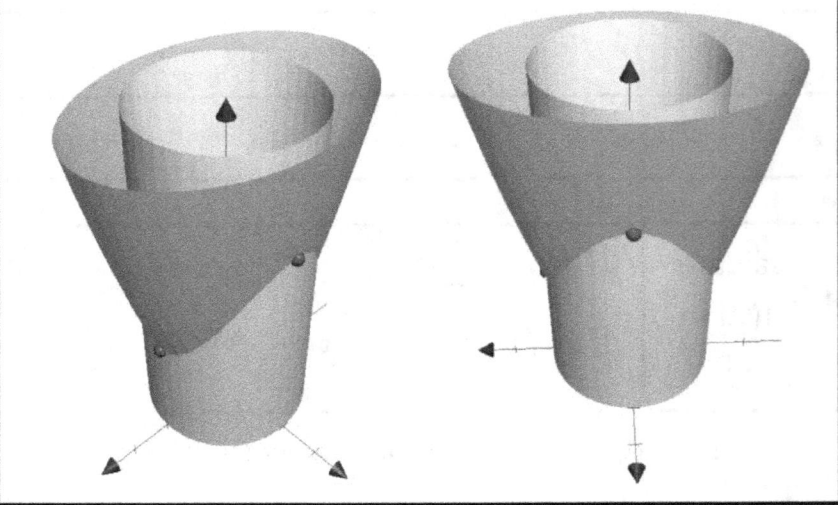

Multiple Integrals

Double Integrals Over Rectangles

Compare Single and Double Integrals

Single Integral: $Area = \int_0^a f(x)\,dx$	
Domain	$0 \leq x \leq a$
Function	$y = f(x)$
Area Diagram	

Double Integral: $Volume = \int_0^a \int_0^b f(x,y)\,dy\,dx$	
Domain	$0 \leq x \leq a$ $0 \leq y \leq b$
Function	$height = z = f(x,y)$
Volume Diagram	

Double Integrals Over Rectangles	
$\int_{x=a}^{x=b} \int_{y=c}^{y=d} f(x,y) \; dy \, dx$	

Do inner integral first	$\int_{x=a}^{x=b} \left[\int_{y=c}^{y=d} f(x,y) \; dy \right] dx$
Do outer integral last	$\int_{x=a}^{x=b} \left[f(x) \right] dx$
Changing order is OK. But, do inner integral first.	$\int_{y=c}^{y=d} \int_{x=a}^{x=b} f(x,y) \; dx \, dy$ $\int_{y=c}^{y=d} \left[\int_{x=a}^{x=b} f(x,y) \; dx \right] dy$

Domain	$a \le x \le b$ $c \le y \le d$

Double Integrals Over Rectangles With Separable Functions
$\int_{x=a}^{x=b} \int_{y=c}^{y=d} g(x)\,h(y)\ dy\,dx$
$\int_{x=a}^{x=b} g(x)\ dx \int_{y=c}^{y=d} h(y)\ dy$

Double Integrals Over Rectangles
Average Value

Recall, for single integrals,

the average value is defined as:

$$f_{ave} = \frac{1}{b-a} \int_a^b f(x) \ dx$$

For double integrals,

the average value is defined as:

$$f_{ave} = \frac{1}{A(R)} \iint_R f(x,y) \ dA$$

Where:

$$R = \text{Rectangle} = [a,b] \times [c,d]$$

$$A(R) = \text{Area of the Rectangle}$$

Double Integrals Over Rectangles -- Ex. 1a

Evaluate the integral.

$$I = \int_0^1 \int_2^3 (3x^2 - 2y) \, dy \, dx$$

$$I = \int_0^1 [\, 3x^2 y - y^2 \,]_2^3 \; dx$$

$$I = \int_0^1 [\, (9x^2 - 9) - (6x^2 - 4) \,] \; dx$$

$$I = \int_0^1 [\, 3x^2 - 5 \,] \; dx$$

$$I = [\, x^3 - 5x \,]_0^1$$

$$I = (1 - 5) - (0 - 0)$$

$$I = -4$$

Double Integrals Over Rectangles -- Ex. 1b

Evaluate the integral.   Switch the order.

$$I = \int_0^1 \int_2^3 (3x^2 - 2y)\, dy\, dx$$

$I = \int_2^3 \int_0^1 (3x^2 - 2y)\, dx\, dy$

$I = \int_2^3 [\, x^3 - 2xy \,]_{x=0}^{x=1}\, dy$

$I = \int_2^3 [(1 - 2y) - (0 - 0)]\, dy$

$I = \int_2^3 [1 - 2y]\, dy$

$I = [\, y - y^2 \,]_{y=2}^{y=3}$

$I = (3 - 9) - (2 - 4)$

$I = -6 + 2 \; = \; -4$ (Same answer)

Double Integrals Over Rectangles -- Ex. 1c

Evaluate the integral.   Separate grouping.

$$I = \int_0^1 \int_2^3 (3x^2 - 2y) \, dy \, dx$$

$I = \int_0^1 3x^2 \, dx \; - \; \int_2^3 2y \, dy$

$I = [x^3]_{x=0}^{x=1} \; - \; [y^2]_{y=2}^{y=3}$

$I = [1 - 0] \; - \; [9 - 4]$

$I = 1 - 5 \;\; = \;\; -4$ (Same answer)

Double Integrals Over Rectangles -- Ex. 2a
Average Value

Let the given function define the height of a mountainous region. If we chop off the top half and use it to fill the bottom half (the valleys), the land will be flat. Find the average height of the region so we know the target height.

Find the average height for the region, 6 square miles, measured from the SW corner.

$$\text{Given:} \quad z = x + xy + 3y^2$$

$$f_{ave} = \frac{1}{A(R)} \iint_R f(x,y) \; dA$$

$$f_{ave} = \frac{1}{36} \int_0^6 \int_0^6 (x + xy + 3y^2) \; dy \, dx$$

$$f_{ave} = \frac{1}{36} \int_0^6 \left[xy + \frac{xy^2}{2} + y^3 \right]_0^6 dx$$

Double Integrals Over Rectangles -- Ex. 2b
Average Value

Find the average height for the region, 6 square miles, measured from the SW corner.

Given: $z = x + xy + 3y^2$

$f_{ave} = \frac{1}{A(R)} \iint_R f(x,y) \, dA$

$f_{ave} = \frac{1}{36} \int_0^6 \int_0^6 (x + xy + 3y^2) \, dy \, dx$

$f_{ave} = \frac{1}{36} \int_0^6 \left[xy + \frac{xy^2}{2} + y^3 \right]_0^6 dx$

$f_{ave} = \frac{1}{36} \int_0^6 [6x + 18x + 216]_0^6 \, dx$

$f_{ave} = \frac{1}{36} \int_0^6 [24x + 216]_0^6 \, dx$

$f_{ave} = \frac{1}{36} [12x^2 + 216x]_0^6$

$f_{ave} = \frac{1}{36} [1728] = 48 \; miles$

Double Integrals Over General Regions

Double Integrals Over General Regions
$$\int_{x=a}^{x=b} \int_{y=g_1(x)}^{y=g_2(x)} f(x,y) \; dy \; dx$$

Do inner integral first	$$\int_{x=a}^{x=b} \left[\int_{y=g_1(x)}^{y=g_2(x)} f(x,y) \; dy \right] dx$$
Do outer integral last	$$\int_{x=a}^{x=b} [f(x)] \; dx$$

Changing order is OK. But, do inner integral first.	$$\int_{y=c}^{y=d} \int_{x=a}^{x=b} f(x,y) \; dx \; dy$$ $$\int_{y=c}^{y=d} \left[\int_{x=h_1(y)}^{x=h_2(y)} f(x,y) \; dx \right] dy$$
Always do outer integral last	$$\int_{y=c}^{y=d} [f(x)] \; dy$$

Note	Boundaries of the inner integral may be functions.

Double Integrals Over General Regions -- Ex. 1a

Evaluate: $\iint_D (1)\ dA$

Where dA is the region bounded by two curves:

$$y = 2x \qquad \text{and} \qquad y = x^2$$

Sketch the Region	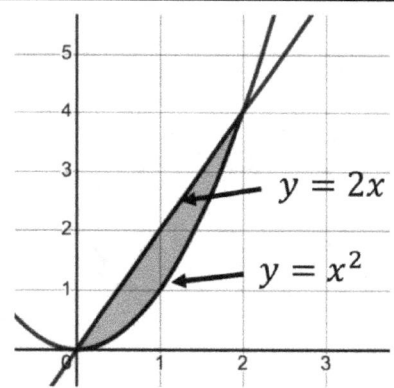
Find Intersections	$(0,0)$ and $(2,4)$

$I \;=\; \iint_D (1)\ dA$

$I \;=\; \int_{x=0}^{x=2} \int_{y=x^2}^{y=2x} (1)\ dy\ dx$

$I \;=\; \int_{x=0}^{x=2} [y]_{y=x^2}^{y=2x}\ dx$

Double Integrals Over General Regions -- Ex. 1b

Evaluate: $\iint_D (1) \, dA$

Where dA is the region bounded by two curves:

$$y = 2x \qquad \text{and} \qquad y = x^2$$

$I = \iint_D (1) \, dA$

$I = \int_{x=0}^{x=2} \int_{y=x^2}^{y=2x} (1) \, dy \, dx$

$I = \int_{x=0}^{x=2} [y]_{y=x^2}^{y=2x} \, dx$

$I = \int_{x=0}^{x=2} [2x - x^2] \, dx$

$I = \left[x^2 - \frac{x^3}{3} \right]_0^2$

$I = 4 - \frac{8}{3} = \frac{4}{3}$

Double Integrals Over General Regions -- Ex. 1c

Evaluate: $\iint_D (1)\ dA$

Where dA is the region bounded by two curves:

$$y = 2x \qquad \text{and} \qquad y = x^2$$

$$I = \iint_D (z)\ dA \ = \ \iint_D (1)\ dA$$

$$I = \int_0^2 \int_{x^2}^{2x} (1)\ dy\ dx \ = \ \frac{4}{3}$$

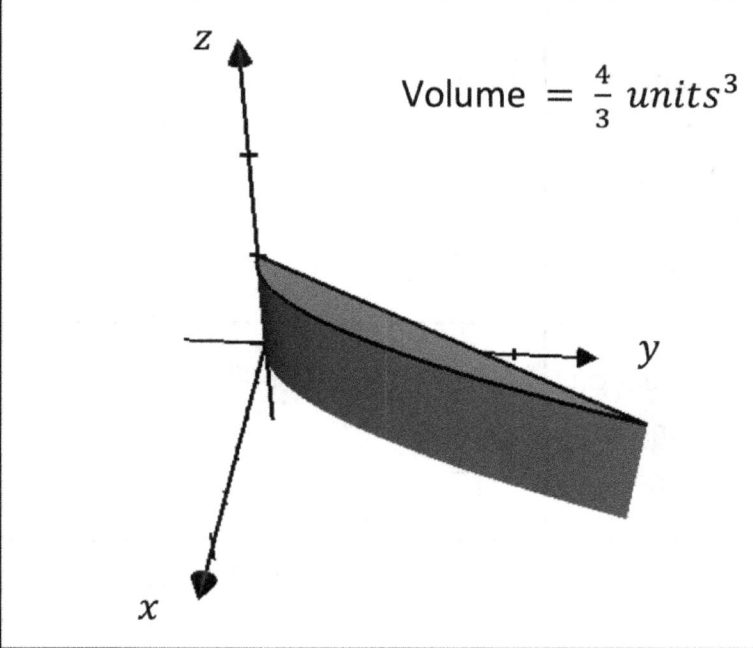

Volume $= \dfrac{4}{3}\ units^3$

Double Integrals Over General Regions -- Ex. 2a

Find the volume of the tetrahedron
bounded by the planes:

$$x + 2y + z = 2 \quad , \quad x = 2y$$

$$x = 0 \quad , \quad y = 0$$

Set $z = 0$ to find where the 1st plane intersects the xy-plane.	$x + 2y + 0 = 2$ $y = \dfrac{2 - x}{2}$ $y = -\dfrac{1}{2}x + 1$
Write the 2nd plane as $y = f(x)$.	$x = 2y$ $y = \dfrac{1}{2}x$
Find intersection of lines in the xy-plane	$\dfrac{1}{2}x = -\dfrac{1}{2}x + 1$ $x = 1 \quad \rightarrow \quad y = \dfrac{1}{2}$

(Stewart, Calculus Early Transcendentals, p. 1005)

Double Integrals Over General Regions -- Ex. 2a

Find the volume of the tetrahedron
bounded by the planes:

$$x + 2y + z = 2 \quad , \quad x = 2y$$

$$x = 0 \quad , \quad y = 0$$

| Previously Found | $y = -\dfrac{1}{2}x + 1$ |
| | $y = \dfrac{1}{2}x$ |

Use above equations to
sketch the domain on the xy-plane.

DOMAIN

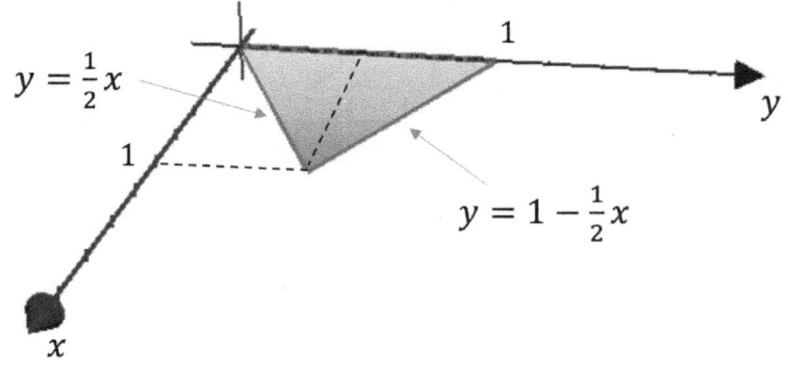

Double Integrals Over General Regions -- Ex. 2c

Find the volume of the tetrahedron
bounded by the planes:

$$x + 2y + z = 2 \quad , \quad x = 2y$$

$$x = 0 \quad , \quad y = 0$$

Previously Found Domain	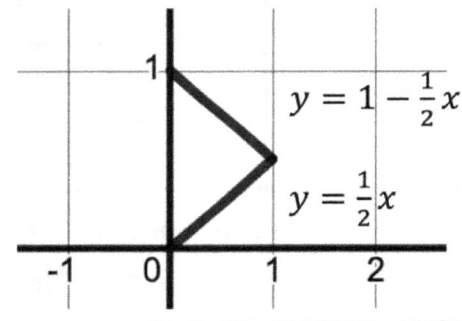

Write the 1st equation to show height

$$\text{height} = z = f(x, y)$$

$$z = 2 - x - 2y$$

$$V = \iint_D (2 - x - 2y) \, dA$$

$$V = \int_0^1 \int_{\frac{1}{2}x}^{1 - \frac{1}{2}x} (2 - x - 2y) \, dy \, dx$$

Double Integrals Over General Regions -- Ex. 2d

Find the volume of the tetrahedron
bounded by the planes:

$$x + 2y + z = 2 \quad , \quad x = 2y$$

$$x = 0 \quad , \quad y = 0$$

Sketch a
3D view

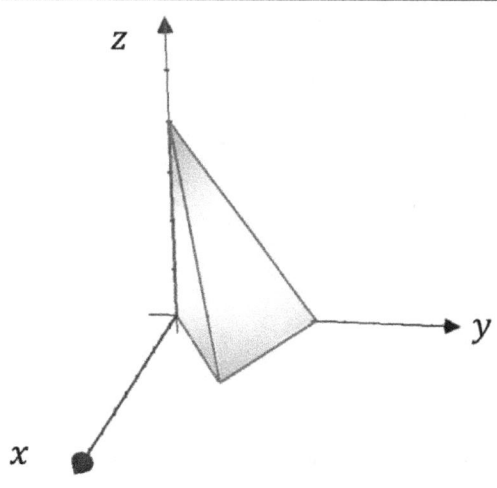

$$V = \int_0^1 \int_{\frac{1}{2}x}^{1-\frac{1}{2}x} (2 - x - 2y) \ dy \ dx$$

$$V = \int_0^1 \left[2y - xy - y^2 \right]_{\frac{1}{2}x}^{1-\frac{1}{2}x} dx$$

Double Integrals Over General Regions -- Ex. 2e

Find the volume of the tetrahedron
bounded by the planes:

$$x + 2y + z = 2 \quad , \quad x = 2y$$

$$x = 0 \quad , \quad y = 0$$

$$V = \int_0^1 [\, 2y - xy - y^2 \,]_{\frac{1}{2}x}^{1-\frac{1}{2}x} \; dx$$

$$V = \int_0^1$$

$$\left[(2 - x) - \left(x - \frac{x^2}{2} \right) - \left(1 - \frac{x}{2} \right)^2 \right]$$

$$- \left[x - \frac{x^2}{2} - \frac{x^2}{4} \right] \; dx$$

$$V = \int_0^1$$

$$\left[2 - 2x + \frac{x^2}{2} - \left(1 - x + \frac{x^2}{4} \right) \right]$$

$$- \left[x - \frac{x^2}{2} - \frac{x^2}{4} \right] \; dx$$

Double Integrals Over General Regions -- Ex. 2f

Find the volume of the tetrahedron
bounded by the planes:

$$x + 2y + z = 2 \quad , \quad x = 2y$$

$$x = 0 \quad , \quad y = 0$$

$$V = \int_0^1$$
$$\left[2 - 2x + \frac{x^2}{2} - 1 + x - \frac{x^2}{4} \right]$$
$$- \left[x - \frac{x^2}{2} - \frac{x^2}{4} \right] \, dx$$

$$V = \int_0^1 \left[1 - 2x + x^2 \right] \, dx$$

$$V = \left[x - x^2 + \frac{x^3}{3} \right]_0^1$$

$$V = \frac{1}{3} \ units^3$$

Double Integrals in Polar Coordinates

Double Integrals – Using Polar Coordinates

When integrating over a circular region, it is easier to define the region in Polar Coordinates. Using rectangular coordinates is possible but complicated.

$$r^2 = x^2 + y^2$$

$$x = r\cos(\theta)$$

$$y = r\sin(\theta)$$

$$\iint_R f(x,y)\ dA \ =$$

$$\int_\alpha^\beta \int_a^b f(r\cos\theta, r\sin\theta)\ r\ dr\ d\theta$$

NOTE: $dA = r\ dr\ d\theta$

Double Integrals – Using Polar Coordinates
$$r = f(\theta)$$

The graph below shows an example of a region,

in **Polar Coordinates**, where the radius is not

constant. Here, $r = f(\theta)$

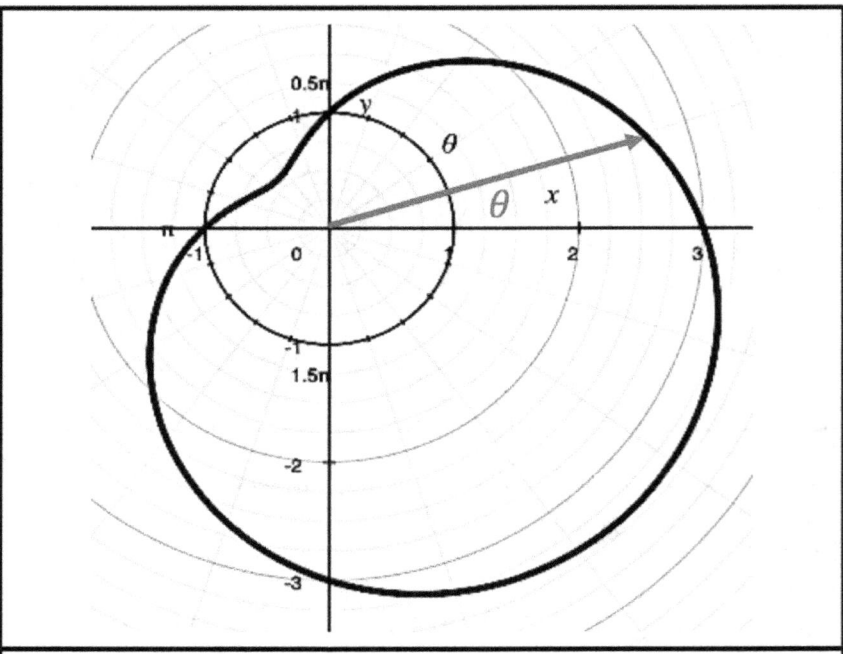

$\iint_R f(x,y) \; dA \; =$

$\int_\alpha^\beta \int_{r \, = \, h_1(\theta)}^{r \, = \, h_2(\theta)} f(r \cos \theta \, , r \sin \theta) \; r \; dr \; d\theta$

Double Integrals – Using Polar Coordinates -- Ex. 1a

Find the volume of the solid, bounded by

the plane: $z = 0$ and

the paraboloid: $z = 4 - x^2 - y^2$

Set $z = 0$ To find where the paraboloid intersects the xy-plane	$0 = 4 - x^2 - y^2$ $x^2 + y^2 = 4$ (circle) Circle with radius $= 2$ Here: $r = 2$

$z = 4 - x^2 - y^2$

$z = 4 - (r \cos \theta)^2 - (r \sin \theta)^2$

$z = 4 - r^2 (\cos^2 \theta + \sin^2 \theta)$

$z = 4 - r^2$

$z = $ height

$V = \iint_D f(r, \theta) \ dA$

$V = \int_{\theta = 0}^{\theta = 2\pi} \int_{r=0}^{r=2} (4 - r^2) \ r \ dr \ d\theta$

Double Integrals – Using Polar Coordinates -- Ex. 1b

Find the volume of the solid, bounded by

the plane: $z = 0$ and

the paraboloid: $z = 4 - x^2 - y^2$

Previously Found	$V = \int_{\theta=0}^{\theta=2\pi} \int_{r=0}^{r=2} (4 - r^2)\, r\, dr\, d\theta$

$V = \int_{\theta=0}^{\theta=2\pi} \int_{r=0}^{r=2} (4r - r^3)\ dr\, d\theta$

$V = \int_{\theta=0}^{\theta=2\pi} \left[2r^2 - \frac{r^4}{4} \right]_{r=0}^{r=2} d\theta$

$V = \int_{\theta=0}^{\theta=2\pi} [(8 - 4) - (0)]\, d\theta$

$V = \int_{\theta=0}^{\theta=2\pi} [4]\, d\theta$

$V = [4\theta]_0^{2\pi} = 8\pi$

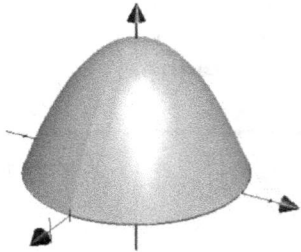

Double Integrals – Using Polar Coordinates -- Ex. 2a

Use a double integral to find the area of one leaf of

the rose: $r = \cos(3\theta)$

Sketch the rose to determine range needed for one leaf.

Use: $-\dfrac{\pi}{3} \le \theta \le \dfrac{\pi}{3}$

And: $0 \le r \le 1$

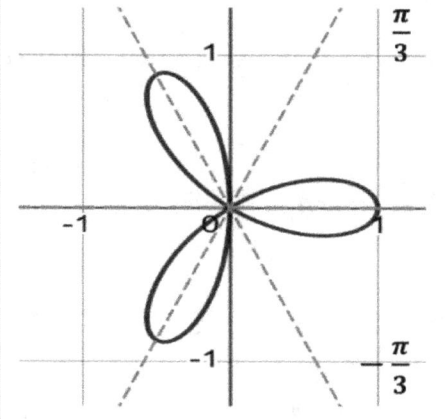

$A = \iint_D f(r, \theta)\ dA$

$A = \int_{\theta = -\frac{\pi}{3}}^{\theta = \frac{\pi}{3}} \int_{r=0}^{r = \cos 3\theta} r\ dr\ d\theta$

$A = \int_{\theta = -\frac{\pi}{3}}^{\theta = \frac{\pi}{3}} \left[\frac{1}{2} r^2 \right]_0^{\cos 3\theta} d\theta$

$A = \frac{1}{2} \int_{\theta = -\frac{\pi}{3}}^{\theta = \frac{\pi}{3}} [\cos^2 3\theta]\ d\theta$

Double Integrals – Using Polar Coordinates -- Ex. 2b

Use a double integral to find the area of one leaf of

the rose: $r = \cos(3\theta)$

$A = \dfrac{1}{2} \displaystyle\int_{\theta = -\frac{\pi}{3}}^{\theta = \frac{\pi}{3}} [\cos^2 3\theta] \; d\theta$

$A = \dfrac{1}{2} \displaystyle\int_{\theta = -\frac{\pi}{3}}^{\theta = \frac{\pi}{3}} \left[\dfrac{1 + \cos(2\cdot3\theta)}{2} \right] \; d\theta$

$A = \dfrac{1}{4} \displaystyle\int_{\theta = -\frac{\pi}{3}}^{\theta = \frac{\pi}{3}} [1 + \cos 6\theta] \; d\theta$

$4A = \displaystyle\int_{-\frac{\pi}{3}}^{\frac{\pi}{3}} [1] \, d\theta \; + \; \int_{-\frac{\pi}{3}}^{\frac{\pi}{3}} [\cos 6\theta] \, d\theta$

$4A = \displaystyle\int_{-\frac{\pi}{3}}^{\frac{\pi}{3}} [1] \, d\theta \; + \; \dfrac{1}{6}\int_{-\frac{\pi}{3}}^{\frac{\pi}{3}} [\cos 6\theta] \, 6 \, d\theta$

$4A = [\,\theta\,]_{-\frac{\pi}{3}}^{\frac{\pi}{3}} \; + \; \dfrac{1}{6}[\,\sin 6\theta\,]_{-\frac{\pi}{3}}^{\frac{\pi}{3}}$

Double Integrals – Using Polar Coordinates -- Ex. 2c

Use a double integral to find the area of one leaf of

the rose: $r = \cos(3\theta)$

$$4A = [\theta]\Big|_{-\frac{\pi}{3}}^{\frac{\pi}{3}} + \frac{1}{6}[\sin 6\theta]\Big|_{-\frac{\pi}{3}}^{\frac{\pi}{3}}$$

$$4A = \left[\frac{2\pi}{3}\right] + \frac{1}{6}[\sin(2\pi) - \sin(-2\pi)]$$

$$4A = \left[\frac{2\pi}{3}\right] + 0$$

$$A = \frac{2\pi}{12} = \frac{\pi}{6}$$

The area of

one leaf is

$\frac{\pi}{6}$ $units^2$

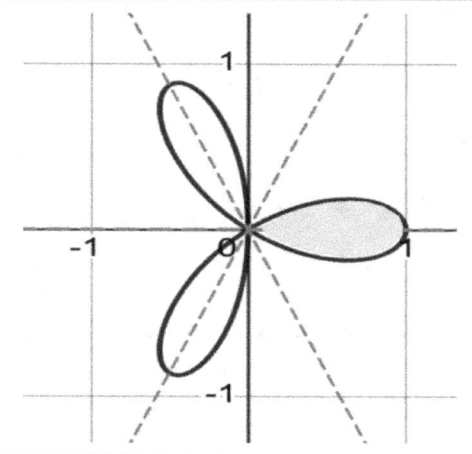

Double Integrals – Using Polar Coordinates -- Ex. 3a

Find the volume of the solid that lies

under the paraboloid: $z = x^2 + y^2$ and

above the x-y Plane   and

inside the cylinder: $x^2 + y^2 = 2x$

Paraboloid intersects the xy-plane when $(z = 0)$

$0 = x^2 + y^2$ $\rightarrow$ $x = y = 0$

The solid lies above the disk

whose boundary is: $x^2 + y^2 = 2x$

$x^2 + y^2 \ = \ 2x$

$(r \cos \theta)^2 + (r \sin \theta)^2 \ = \ 2 r \cos \theta$

$r^2 (\cos \theta + \sin \theta) \ = \ 2r \cos \theta$

$r^2 (1) \ = \ 2r \cos \theta$

$r \ = \ 2 \cos \theta$ (Boundary of disk)

(Stewart, Calculus Early Transcendentals, p. 1014)

Double Integrals – Using Polar Coordinates -- Ex. 3b

Find the volume of the solid that lies

under the paraboloid: $z = x^2 + y^2$ and

above the x-y Plane   and

inside the cylinder: $x^2 + y^2 = 2x$

Previously found boundary of disk	$r = 2\cos\theta$
Make a sketch of the boundary on the. xy-plane.	
$\theta = -\dfrac{\pi}{2} \quad \rightarrow r = 0$ $\theta = \ \ 0 \quad \rightarrow r = 2$ $\theta = \ \ \dfrac{\pi}{2} \quad \rightarrow r = 0$	One Loop: $0 \ \le \ r \ \le \ 2\cos\theta$ $-\dfrac{\pi}{2} \ \le \ \theta \ \le \ \dfrac{\pi}{2}$

Double Integrals – Using Polar Coordinates -- Ex. 3c

Find the volume of the solid that lies

under the paraboloid: $z = x^2 + y^2$ and

above the x-y Plane   and

inside the cylinder: $x^2 + y^2 = 2x$

$0 \leq r \leq 2 \cos \theta$

$-\dfrac{\pi}{2} \leq \theta \leq \dfrac{\pi}{2}$

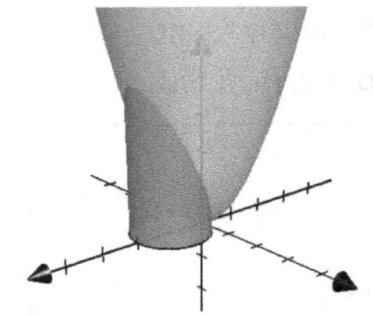

$V = \iint_D f(r, \theta) \ dA$

$V = \int_{\theta = -\frac{\pi}{2}}^{\theta = \frac{\pi}{2}} \int_{r=0}^{r=2\cos\theta} (z) \ r \ dr \ d\theta$

$V = \int_{-\frac{\pi}{2}}^{\frac{\pi}{2}} \int_{0}^{2\cos\theta} (x^2 + y^2) \ r \ dr \ d\theta$

$V = \int_{-\frac{\pi}{2}}^{\frac{\pi}{2}} \int_{0}^{2\cos\theta} (r^2) \ r \ dr \ d\theta$

Double Integrals – Using Polar Coordinates -- Ex. 3d

Continued ...

$$V = \int_{-\frac{\pi}{2}}^{\frac{\pi}{2}} \int_0^{2\cos\theta} (r^2)\ r\ dr\ d\theta$$

$$V = \int_{-\frac{\pi}{2}}^{\frac{\pi}{2}} \int_0^{2\cos\theta} (r^3)\ dr\ d\theta$$

$$V = \int_{-\frac{\pi}{2}}^{\frac{\pi}{2}} \left[\ \frac{r^4}{4}\ \right]_0^{2\cos\theta}\ d\theta$$

$$V = \frac{1}{4} \int_{-\frac{\pi}{2}}^{\frac{\pi}{2}} [\ r^4\ ]_0^{2\cos\theta}\ d\theta$$

$$V = \frac{1}{4} \int_{-\frac{\pi}{2}}^{\frac{\pi}{2}} [\ 16\cos^4\theta\ -\ 0\ ]\ d\theta$$

$$V = 4 \int_{-\frac{\pi}{2}}^{\frac{\pi}{2}} [\cos^4\theta\ ]\ d\theta$$

$$\frac{V}{4} = \int_{-\frac{\pi}{2}}^{\frac{\pi}{2}} [\cos^4\theta\ ]\ d\theta$$

$$\frac{V}{4} = \int_{-\frac{\pi}{2}}^{\frac{\pi}{2}} [\cos^2\theta\ ]^2\ d\theta$$

Double Integrals – Using Polar Coordinates -- Ex. 3e

Continued ...

$$\frac{V}{4} = \int_{-\frac{\pi}{2}}^{\frac{\pi}{2}} [\cos^2 \theta]^2 \, d\theta$$

$$\frac{V}{4} = \int_{-\frac{\pi}{2}}^{\frac{\pi}{2}} \left[\frac{1 + \cos 2\theta}{2} \right]^2 \, d\theta$$

$$\frac{V}{4} = \frac{1}{4} \int_{-\frac{\pi}{2}}^{\frac{\pi}{2}} [1 + 2\cos 2\theta + \cos^2 2\theta] \, d\theta$$

$$V = \int_{-\frac{\pi}{2}}^{\frac{\pi}{2}} [1 + 2\cos 2\theta + \cos^2 2\theta] \, d\theta$$

$$V = 2\int_{0}^{\frac{\pi}{2}} [1 + 2\cos 2\theta + \cos^2 2\theta] \, d\theta$$

$$\frac{V}{2} = \int_{0}^{\frac{\pi}{2}} [1 + 2\cos 2\theta + \cos^2 2\theta] \, d\theta$$

$$\frac{V}{2} = I_1 + I_2 + I_3$$

Double Integrals – Using Polar Coordinates -- Ex. 3f

$$\frac{V}{2} = I_1 + I_2 + I_3$$

$$I_1 = \int_0^{\frac{\pi}{2}} [\, 1 \,] \, d\theta \;\; = \;\; [\theta]_0^{\frac{\pi}{2}} \;\; = \;\; \frac{\pi}{2}$$

$$I_2 = \int_0^{\frac{\pi}{2}} [\, 2\cos 2\theta \,] \, d\theta$$

$$I_2 = \; 2[\, \sin 2\theta \,]_0^{\frac{\pi}{2}} \;\; = \;\; 2\,[\sin \pi] \;\; = \; 0$$

$$I_3 \; = \; \int_0^{\frac{\pi}{2}} [\, \cos^2 2\theta \,] \; d\theta$$

$$I_3 \; = \int_0^{\frac{\pi}{2}} \left[\frac{1 + \cos 4\theta}{2} \right] \; d\theta$$

$$I_3 \; = \; \frac{1}{2} \int_0^{\frac{\pi}{2}} [\, 1 \, + \, \cos 4\theta \,] \; d\theta$$

$$I_3 = \; \frac{1}{2} \int_0^{\frac{\pi}{2}} [\, 1 \,] \; d\theta \; + \; \frac{1}{8} \int_0^{\frac{\pi}{2}} [\, \cos 4\theta \,] \; 4 \, d\theta$$

$$I_3 = \; \frac{1}{2} [\, \theta \,]_0^{\frac{\pi}{2}} \; + \; \frac{1}{8} [\sin 4\theta \,]_0^{\frac{\pi}{2}} \;\; = \;\; \frac{1}{2}\left[\frac{\pi}{2} \right] \;\; = \;\; \frac{\pi}{4}$$

Double Integrals – Using Polar Coordinates -- Ex. 3g

Find the volume of the solid that lies

under the paraboloid: $z = x^2 + y^2$ and

above the x-y Plane   and

inside the cylinder: $x^2 + y^2 = 2x$

$$\frac{V}{2} = I_1 + I_2 + I_3$$

$$\frac{V}{2} = \frac{\pi}{2} + 0 + \frac{\pi}{4} = \frac{3\pi}{4}$$

$$V = \frac{3\pi}{2}$$

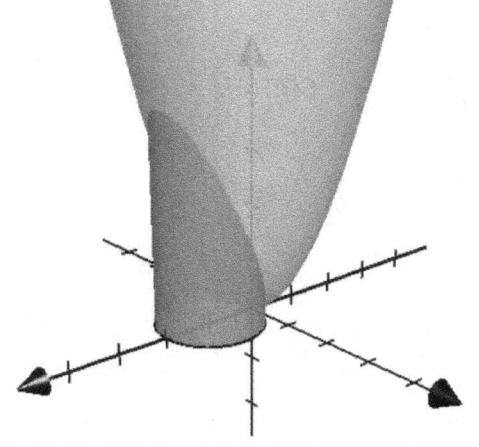

Applications of Double Integrals

Mass and Density Function

Mass and Density
$m = mass$ $m = density \cdot area$
$m = \iint_D \rho(x,y) \; dA$ $\rho(x,y) =$ Mass density function

Density Function
In general, the total amount of an item, over an area can be calculated by taking a double integral of the density function for that item, over the area. $\text{Total Amount} = \iint_D \rho(x,y) \; dA$
$\rho(x,y) =$ Density function for the item

Density Function -- Ex. 1a

Find the total charge, Q, over the area shown.

The charge density function is:

$\rho(x, y) = xy$

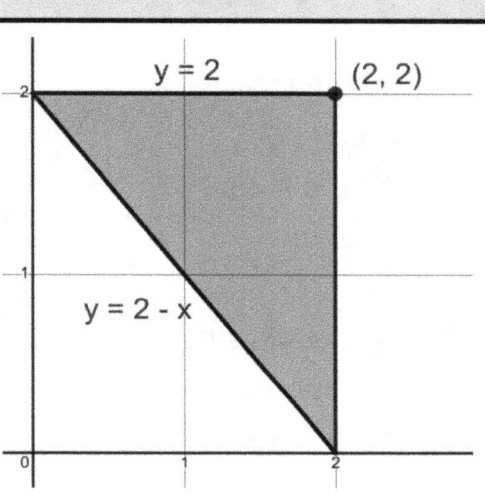

$y = 2$ (2, 2)

$y = 2 - x$

$Q = \iint_D \rho(x, y) \; dA$

$Q = \int_0^2 \int_{y=2-x}^{y=2} (xy) \; dy \, dx$

$Q = \int_0^2 \left[x \, \frac{y^2}{2} \right]_{y=2-x}^{y=2} \; dx$

$Q = \frac{1}{2} \int_0^2 [xy^2]_{y=2-x}^{y=2} \; dx$

$Q = \frac{1}{2} \int_0^2 x[4 - (2-x)^2] \; dx$

$Q = \frac{1}{2} \int_0^2 x[4 - (4 - 4x + x^2)] \; dx$

Density Function -- Ex. 1b

Find the total charge, Q, over the area shown.

The charge density function is:

$$\rho(x, y) = xy$$

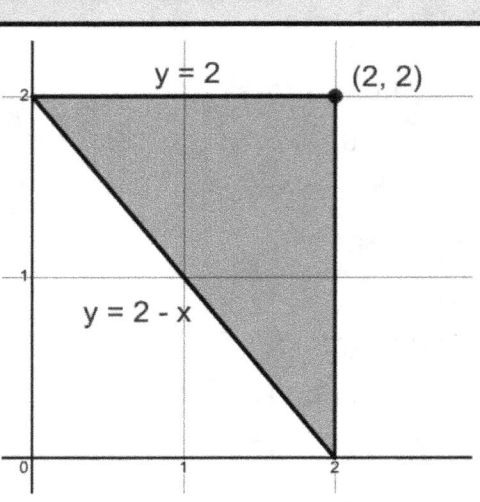

$$Q = \frac{1}{2} \int_0^2 x[\,4x - x^2\,]\ dx$$

$$Q = \frac{1}{2} \int_0^2 4x^2 - x^3\ dx$$

$$Q = \frac{1}{2} \left[\frac{4\,x^3}{3} - \frac{x^4}{4} \right]_0^2$$

$$Q = \frac{1}{2} \left[\frac{4\,(8)}{3} - \frac{(16)}{4} \right]$$

$$Q = \frac{1}{2} \left[\frac{32}{3} - 4 \right] = \frac{1}{2} \left[\frac{20}{3} \right] = \frac{10}{3}$$

$$Q = \frac{10}{3} \; \frac{C}{m^2}$$

Density Function -- Ex. 2a

Find the total charge, Q, over the area shown.

The charge density function is:

$\rho(x, y) = xy$

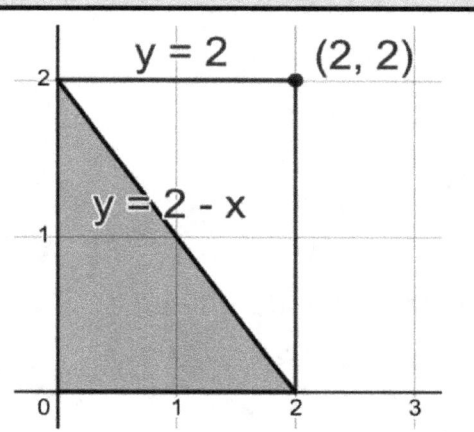

$$Q = \iint_D \rho(x, y) \; dA$$

$$Q = \int_0^2 \int_{y=0}^{y=2-x} (xy) \; dy \, dx$$

$$Q = \int_0^2 \left[x \frac{y^2}{2} \right]_{y=0}^{y=2-x} dx$$

$$Q = \frac{1}{2} \int_0^2 \left[xy^2 \right]_{y=0}^{y=2-x} dx$$

$$Q = \frac{1}{2} \int_0^2 x(2-x)^2 \; dx$$

$$Q = \frac{1}{2} \int_0^2 x[4 - 4x + x^2] \; dx$$

Density Function -- Ex. 2b

Find the total charge, Q, over the area shown.

The charge density function is:

$$\rho(x, y) = xy$$

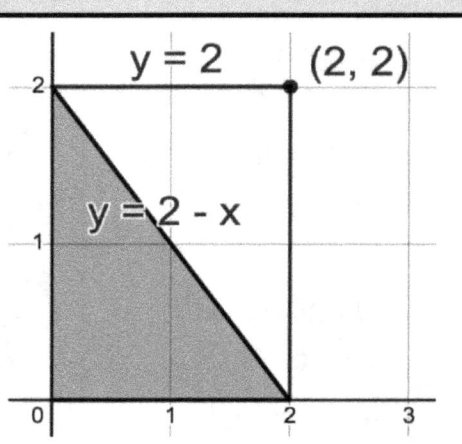

$y = 2$

$(2, 2)$

$y = 2 - x$

$$Q = \frac{1}{2} \int_0^2 x [4 - 4x + x^2] \ dx$$

$$Q = \frac{1}{2} \int_0^2 [4x - 4x^2 + x^3] \ dx$$

$$Q = \frac{1}{2} \left[2x^2 - \frac{4x^3}{3} + \frac{x^4}{4} \right]_0^2$$

$$Q = \frac{1}{2} \left[8 - \frac{32}{3} + \frac{16}{4} \right] = \frac{1}{2} \left[\frac{4}{3} \right] = \frac{2}{3}$$

$$Q = \frac{2}{3} \ \frac{C}{m^2}$$

Density Function -- Ex. 3a

Find the total charge, Q, over the area shown.

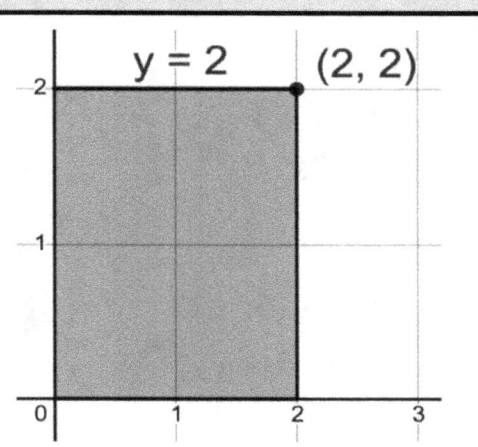

The charge density function is:
$$\rho(x, y) = xy$$

$Q = \iint_D \rho(x, y) \; dA$

$Q = \int_0^2 \int_{y=0}^{y=2} (xy) \; dy \; dx$

$Q = \int_0^2 \left[x \frac{y^2}{2} \right]_{y=0}^{y=2} dx$

$Q = \frac{1}{2} \int_0^2 \left[xy^2 \right]_{y=0}^{y=2} dx$

$Q = \frac{1}{2} \int_0^2 x \, (2)^2 \; dx$

$Q = \frac{1}{2} \int_0^2 4x \; dx$

Density Function -- Ex. 3b

Find the total
charge, Q,
over the area
shown.

The charge density
function is:
$\rho(x, y) = xy$

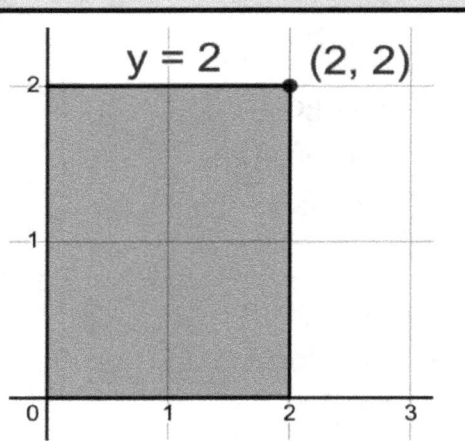

$y = 2$ (2, 2)

$Q = \frac{1}{2} \int_0^2 4x \ dx$

$Q = 2 \int_0^2 x \ dx = 2 \left[\frac{x^2}{2} \right]_0^2 = 2 \left[\frac{4}{2} \right]$

$Q = 4 \ C/m^2$

NOTE: The average charge for this area is the sum of
the average charges for the two areas, calculated in
the previous two examples.

$$\frac{10}{3} + \frac{2}{3} = \frac{12}{3} = 4$$

Moments and Center of Mass

Moments and Center of Mass
M_x = Moment about x-axis $M_x = \iint_D y \cdot \rho(x,y) \ dA$
M_y = Moment about y-axis $M_y = \iint_D x \cdot \rho(x,y) \ dA$
$(\bar{x}, \bar{y})$ = Coordinates of Center of Mass
$\bar{x} = \dfrac{M_y}{m} = \dfrac{1}{m} \iint_D x \cdot \rho(x,y) \ dA$
$\bar{y} = \dfrac{M_x}{m} = \dfrac{1}{m} \iint_D y \cdot \rho(x,y) \ dA$

$$m = mass = \iint_D \rho(x,y) \ dA$$

$$\rho(x,y) = \text{Mass density function}$$

Moments and Center of Mass -- Ex. 1a

Find the mass and center of mass for the triangle with vertices at:

$(0, 2), (2, 2), (2, 0)$

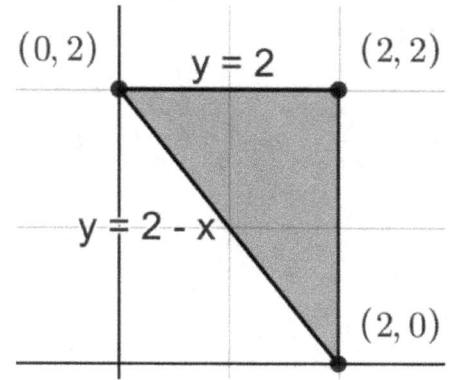

$(0, 2)$ $y = 2$ $(2, 2)$

$y = 2 - x$

$(2, 0)$

The mass density function is: $\rho(x, y) = 1 + 2x + y$

$$m = \iint_D \rho(x, y) \ dA$$

$$m = \int_0^2 \int_{y=2-x}^{y=2} (1 + 2x + y) \ dy \, dx$$

$$m = \int_0^2 \left[y + 2xy + \frac{y^2}{2} \right]_{y=2-x}^{y=2} dx$$

$$m = \int_0^2 \left(\frac{3}{2} x^2 + 3x \right) \ dx$$

$$m = \left[\frac{x^3}{2} + \frac{3}{2} x^2 \right]_0^2 = \left[\frac{8}{2} + \frac{12}{2} \right] = 10$$

Moments and Center of Mass Ex. 1b

Previously found: $m = 10$

$$\bar{x} = \frac{1}{m} \iint_D x \cdot \rho(x,y) \; dA$$

$$\bar{x} = \frac{1}{10} \int_0^2 \int_{y=2-x}^{y=2} x(1 + 2x + y) \; dy \, dx$$

$$\bar{x} = \frac{1}{10} \int_0^2 \int_{2-x}^2 (x + 2x^2 + xy) \; dy \, dx$$

$$\bar{x} = \frac{1}{10} \int_0^2 \left[xy + 2yx^2 + \frac{xy^2}{2} \right]_{2-x}^2 dx$$

$$\bar{x} = \frac{1}{10} \int_0^2 \left(\frac{3x^3}{2} + 3x^2 \right) dx$$

$$\bar{x} = \frac{1}{10} \left[\frac{3x^4}{8} + x^3 \right]_0^2$$

$$\bar{x} = \frac{1}{10} [6 + 8] = \frac{1}{10} [14] = \frac{7}{5}$$

Moments and Center of Mass Ex. 1c

Previously Found	$m = 10$
	$\bar{x} = \dfrac{7}{5} \approx 1.4$

$$\bar{y} = \frac{1}{m} \iint_D y \cdot \rho(x,y) \ dA$$

$$\bar{y} = \frac{1}{10} \int_0^2 \int_{y=2-x}^{y=2} y(1 + 2x + y) \ dy \ dx$$

Desmos can be used to evaluate integrals. $\bar{y} = 1.3$

$$\frac{1}{10} \int_0^2 \int_{2-x}^2 x(1 + 2x + y) \ dy \ dx$$

$$= 1.4$$

$$\frac{1}{10} \int_0^2 \int_{2-x}^2 y(1 + 2x + y) \ dy \ dx$$

$$= 1.33333333333$$

Center of Mass $(\bar{x}, \bar{y}) \approx (1.4, 1.3)$	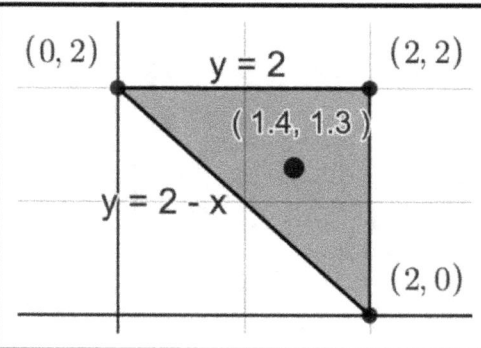

Moments and Center of Mass -- Ex. 2a

Find the center of
mass for the region
bounded by:

$$y = x^2 \quad \& \quad y = 4$$

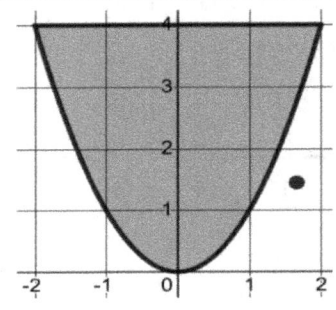

Mass density function is: $\quad \rho(x,y) = 1 + 2y + 6x^2$

$$m = \iint_D \rho(x,y) \ dA$$

$$m = \int_{-2}^{2} \int_{y=0}^{y=x^2} (1 + 2y + 6x^2) \ dy \ dx$$

$$m = 2 \int_0^2 [1 + y^2 + 6x^2 y]_{y=0}^{y=x^2} \ dx$$

$$m = 2 \int_0^2 (x^2 + x^4 + 6x^4) \ dx$$

$$m = 2 \int_0^2 (x^2 + 7x^4) \ dx$$

$$m = 2 \left[\frac{1}{3}x^2 + \frac{7}{5}x^5 \right]_0^2 = \frac{1424}{15} \approx 94.9$$

Moments and Center of Mass -- Ex. 2b

Find the center of
mass for the region
bounded by:

$$y = x^2 \quad \& \quad y = 4$$

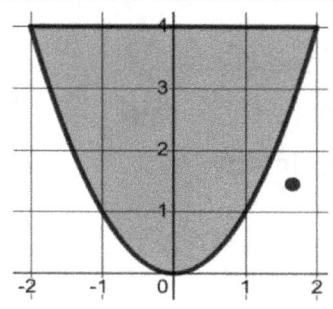

Mass density function is: $\quad \rho(x, y) = 1 + 2y + 6x^2$

Previously Found	$m = \dfrac{1424}{15}$
Use these equations	Center of Mass $= (\bar{x}, \bar{y})$
	$\bar{x} = \dfrac{M_x}{m}, \quad \bar{y} = \dfrac{M_y}{m}$
	$M_x = \iint_D y \, \rho(x, y) \; dA$
	$M_y = \iint_D x \, \rho(x, y) \; dA$
Next	Find: M_x and M_y

Moments and Center of Mass -- Ex. 2c

Find the center of mass for the region bounded by:

$$y = x^2 \ \& \ y = 4$$

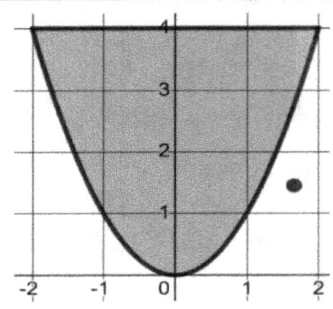

Mass density function is: $\rho(x,y) = 1 + 2y + 6x^2$

| Next | Find: M_x and M_y |

$$M_x = \iint_D y\,\rho(x,y)\,dA$$

$$M_x = 2\int_0^2 \int_0^{x^2} y\,(1 + 2y + 6x^2)\,dy\,dx$$

$$M_x = 2\int_0^2 \left[\frac{1}{2}y^2 + \frac{2}{3}y^3 + \frac{6}{2}y^2x^2\right]_0^{x^2}\,dx$$

$$M_x = 2\int_0^2 x\left[\frac{1}{2}x^4 + \frac{2}{3}x^6 + 3x^6\right]\,dx$$

$$M_x = 2\int_0^2 \left[\frac{1}{2}x^4 + \frac{11}{3}x^6\right]\,dx$$

$$M_x = 2\left[\frac{1}{10}x^5 + \frac{11}{21}x^7\right]_0^2 = 2\left[\frac{7376}{105}\right]$$

Moments and Center of Mass -- Ex. 2d

Find the center of
mass for the region
bounded by:

$$y = x^2 \quad \& \quad y = 4$$

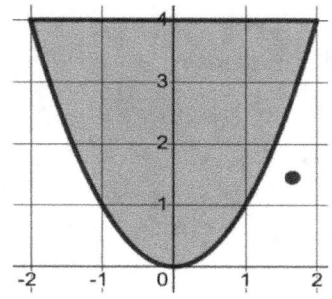

Mass density function is: $\quad \rho(x, y) = 1 + 2y + 6x^2$

Next	Find: M_y

$$M_y \;=\; \iint_D x\,\rho(x, y)\; dA$$

$$M_y \;=\; 2\int_0^2 \int_0^{x^2} x\,(1 + 2y + 6x^2)\; dy\, dx$$

$$M_y \;=\; 2\int_0^2 x\,[\,y + y^2 + 6yx^2\,]_0^{x^2}\; dx$$

$$M_y \;=\; 2\int_0^2 x\,[\,x^2 + x^4 + 6x^4\,]\; dx$$

$$M_y \;=\; 2\int_0^2 [\,x^3 + 7x^5\,]\; dx$$

$$M_y \;=\; 2\left[\tfrac{1}{4}x^4 + \tfrac{7}{6}x^6\right]_0^2 \;=\; \frac{472}{3}$$

Moments and Center of Mass -- Ex. 2e

Find the center of mass for the region bounded by:

$$y = x^2 \quad \& \quad y = 4$$

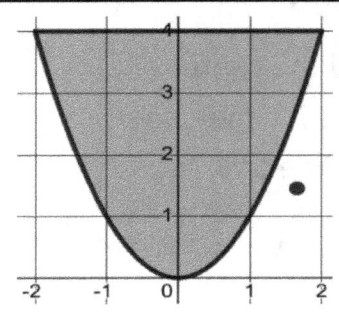

Mass density function is: $\rho(x, y) = 1 + 2y + 6x^2$

Previously Found	$m = \dfrac{1424}{15}$ $M_x = 2\left[\dfrac{7376}{105}\right]$ $M_y = \dfrac{472}{3}$
Find $\bar{x}$ and $\bar{y}$	$\bar{x} = \dfrac{M_x}{m} = \dfrac{2(7376)(15)}{(105)(1424)} = \dfrac{295}{178}$ $\bar{y} = \dfrac{M_y}{m} = \dfrac{(472)(15)}{(3)(1424)} = \dfrac{922}{623}$
Center of Mass	$(\bar{x}, \bar{y}) = $ Center of Mass $(\bar{x}, \bar{y}) = \left(\dfrac{295}{178}, \dfrac{922}{623}\right) \approx (1.7, 1.5)$

Moment of Inertia

Moment of Inertia	
I_x	I_x = Moment of Inertia about x-axis $I_x = \iint_D y^2 \cdot \rho(x, y) \ dA$
I_y	I_y = Moment of Inertia about y-axis $I_y = \iint_D x^2 \cdot \rho(x, y) \ dA$
I_o	I_o = Moment of Inertia about Origin I_o = Polar Moment of Inertia $I_o = \iint_D (x^2 + y^2) \cdot \rho(x, y) \ dA$

$\rho(x, y)$	$\rho(x, y)$ = density function

Moment of Inertia -- Ex. 1a

Find the moments of inertia (I_x, I_y, I_0)

With a constant density function: $\rho(x, y) = k$

(Homogeneous over disk D)

Boundary of D is a circle: $x^2 + y^2 = 9$ $(r = 3)$

Convert to Polar	$I_0 = \iint_D (x^2 + y^2) \cdot \rho(x, y)\ dA$
	$I_0 = \int_0^{2\pi} \int_0^3 r^2 \cdot k \quad r\ dr\ d\theta$
	$I_0 = k \int_0^{2\pi} \int_0^3 r^3 \quad dr\ d\theta$

I_0 = Moment of Inertia about the origin

$$I_0 = \frac{k}{4} \int_0^{2\pi} [r^4]_0^3\ d\theta$$

$$I_0 = \frac{k}{4} \int_0^{2\pi} [3^4]\ d\theta$$

$$I_0 = \frac{81 \cdot k}{4} [\theta]_0^{2\pi}$$

$$I_0 = \frac{81 \cdot k}{4} [2\pi] = \frac{81\, k\, \pi}{2}$$

Moment of Inertia -- Ex. 1b

Find the moments of inertia (I_x, I_y, I_0)

With a constant density function: $\rho(x, y) = k$

(Homogeneous over disk D)

Boundary of D is a circle: $x^2 + y^2 = 9$ $(r = 3)$

Previously Found	$I_0 = \dfrac{81\, k\, \pi}{2}$
Because of symmetry and homogeneous density function ...	$I_x + I_y = I_0$ $I_x = I_y$
$I_x =$ Moment of Inertia about x-axis	$I_x + I_y = I_0$ $I_x + I_x = I_0$ $2\, I_x = \dfrac{81\, k\, \pi}{2}$ $I_x = \dfrac{81\, k\, \pi}{4}$
Find I_y	$I_y = I_x = \dfrac{81\, k\, \pi}{4}$

Probability and Expected Value

Probability
$P(a \le X \le b) \;=\; \int_a^b f(x)\,dx$
$\big((X,Y) \in D\big) \;=\; \iint_D f(x,y)\,dA$
$P\,(a \le X \le b,\; c \le Y \le d) \;=\; \int_a^b \int_c^d f(x,y)\,dy\,dx$
$0 \le P \le 1$

Expected Value	
mean	$\mu \;=\; \text{mean} \;=\; \int_{-\infty}^{\infty} x\,f(x)\,dx$
X-mean	$X\text{-mean} \;=\; u_1$ $X\text{-mean} \;=\; \iint_{R^2} x \cdot f(x,y)\,dA$
Y-mean	$Y\text{-mean} \;=\; u_2$ $Y\text{-mean} \;=\; \iint_{R^2} y \cdot f(x,y)\,dA$
X and Y	X and Y are random variables with joint density function f .

Probability -- Ex. 1a

If the joint density function for X and Y is:

$$f(x,y) = \begin{cases} C\,(x+2y) & 0 \le x \le 10\,, 0 \le y \le 10 \\ 0 & Otherwise \end{cases}$$

Find C by ensuring the double integral $= 1$.
When outside the 10 x 10 rectangle, $f = 0$
Then, find $P(X \le 7, Y \ge 2)$

$$\int_{-\infty}^{\infty} \int_{-\infty}^{\infty} f(x,y)\; dy\, dx \;=\; \int_0^{10} \int_0^{10} f(x,y)\; dy\, dx$$

$$1 \;=\; \int_0^{10} \int_0^{10} f(x,y)\; dy\, dx$$

$$1 \;=\; \int_0^{10} \int_0^{10} C\,(x+2y)\; dy\, dx$$

$$1 \;=\; C \int_0^{10} [\,xy + y^2\,]_0^{10}\; dx$$

$$1 \;=\; C \int_0^{10} [\,10x + 100\,]\; dx$$

$$1 \;=\; C\,[\,5x^2 + 100x\,]_0^{10}$$

(Stewart, Calculus Early Transcendentals, p. 1022)

Probability -- Ex. 1b

If the joint density function for X and Y is:

$$f(x,y) = \begin{cases} C\,(x + 2y) & 0 \le x \le 10, 0 \le y \le 10 \\ 0 & Otherwise \end{cases}$$

Find C by ensuring the double integral = 1.
When outside the 10 x 10 rectangle, $f = 0$
Then, find $P(X \le 7, Y \ge 2)$

$$1 = C\,[\,5x^2 + 100x\,]_0^{10}$$

$$1 = C\,[\,500 + 1000\,] \quad = \quad 1500\,C$$

$$C = \frac{1}{1500}$$

$$P\,(X \le 7, Y \ge 2) = \int_{-\infty}^{7} \int_{2}^{\infty} f(x,y)\ dy\,dx$$

X is at least 0. Y is at most 10.   So, change infinite bounds.

$$P\,(X \le 7, Y \ge 2) = \int_{0}^{7} \int_{2}^{10} \frac{1}{1500}\,(x + 2y)\ dy\,dx$$

$$P\,(X \le 7, Y \ge 2) = \frac{1}{1500} \int_{0}^{7} [\,xy + y^2\,]_{y=2}^{y=10}\ dx$$

$$= \frac{1}{1500} \int_{0}^{7} [\,10x + 100\,] - [\,2x + 4\,]\ dx$$

Probability -- Ex. 1c

If the joint density function for X and Y is:

$$f(x,y) = \begin{cases} C\,(x+2y) & 0 \le x \le 10\,, 0 \le y \le 10 \\ 0 & Otherwise \end{cases}$$

Find C by ensuring the double integral = 1.

When outside the 10 x 10 rectangle, $f = 0$

Then, find $P(\,X \le 7, Y \ge 2\,)$

Previously Found	$C = \dfrac{1}{1500}$

Continued ...

$$P\,(X \le 7, Y \ge 2) \;=\; \frac{1}{1500}\int_0^7 [\,8x + 96\,]\,dx$$

$$P\,(X \le 7, Y \ge 2) \;=\; \frac{1}{1500}\,[\,4x^2 + 96x\,]_{x=0}^{x=7}$$

$$P\,(X \le 7, Y \ge 2) \;=\; \frac{1}{1500}\,[\,196 + 672\,]$$

$$P\,(X \le 7, Y \ge 2) \;=\; \frac{868}{1500} \;=\; \frac{217}{375} \;\approx\; 0.58$$

Probability -- Ex. 2a

A factory produces (cylindrically shaped) roller bearings that are sold as having a diameter of 4.0 cm and a length of 6.0 cm. The diameters, X, are normally distributed, with mean of 4.0 cm & std. deviation 0.01. The lengths, Y, are normally distributed, with mean of 6.0 cm and standard deviation of 0.01 cm.

Find the probability that a randomly selected roller bearing that has either the diameter or length that differs from the mean by more than 0.02 cm.

Given: The probability distribution function for a normal distribution is:

$$f(x) \; = \; \frac{1}{\sigma \sqrt{2\pi}} \; e^{-\frac{(x-\mu)^2}{2\sigma^2}}$$

Where: $\mu = $ mean.   And $\sigma = $ std. dev.

(Stewart, Calculus Early Transcendentals, p. 1024)

Probability -- Ex. 2b

Given: X and Y are normally distributed.

$$\mu_1 = 4.0 \qquad \mu_2 = 6.0 \qquad \sigma_1 = \sigma_2 = 0.01$$

$$f_1(x) = \frac{1}{0.01\sqrt{2\pi}}\, e^{-\frac{(x-4)^2}{.0002}}$$

$$f_2(y) = \frac{1}{0.01\sqrt{2\pi}}\, e^{-\frac{(y-6)^2}{.0002}}$$

Since X and Y are independent,

$$f(x,y) = f_1(x) \cdot f_2(y)$$

$$f(x,y) = \frac{1}{0.0001\,(2\pi)}\, e^{-\frac{(x-4)^2}{.0002}}\, e^{-\frac{(x-6)^2}{.0002}}$$

$$f(x,y) = \frac{1}{0.0002\,\pi}\, e^{-\frac{(x-4)^2 - (x-6)^2}{.0002}}$$

$$f(x,y) = \frac{5000}{\pi}\, e^{-5000\left[(x-4)^2 - (x-6)^2\right]}$$

Probability -- Ex. 2c

Previously found: Joint density function

$$f(x,y) = \frac{5000}{\pi} e^{-5000[(x-4)^2 - (y-6)^2]}$$

Probability that both X and Y differ from their

means by **LESS** than 0.02 cm

$P(3.98 < X < 4.02, \ 5.98 < Y < 6.02) =$

$$= \int_{3.98}^{4.02} \int_{5.98}^{6.02} f(x,y) \ dy \ dx$$

$$= \frac{5000}{\pi} \int_{3.98}^{4.02} \int_{5.98}^{6.02} e^{-5000[(x-4)^2 - (y-6)^2]} \ dy \ dx$$

Using Desmos to evaluate the integral

as the product of two separate integrals ...

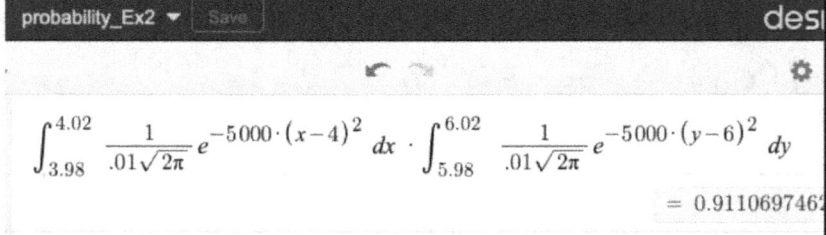

probability_Ex2 ▾ Save   desm

$$\int_{3.98}^{4.02} \frac{1}{.01\sqrt{2\pi}} e^{-5000\cdot(x-4)^2} \ dx \ \cdot \ \int_{5.98}^{6.02} \frac{1}{.01\sqrt{2\pi}} e^{-5000\cdot(y-6)^2} \ dy$$

$$= 0.9110697462$$

Probability -- Ex. 2d

Previously found:

Probability that both X and Y differ from their means by **LESS** than 0.02 cm

$$P(\,3.98 < X < 4.02,\ 5.98 < Y < 6.02\,) \ = \ 0.91$$

Note: Integral was evaluated with Desmos.

Probability that both X and Y differ from their means by **MORE** than 0.02 cm

$$1 - 0.91 \ = \ 0.09$$

<u>Surface Area</u>

Surface Area

$$z \;=\; f(x,y) \;=\; surface$$

$A(S) \;=\;$ Area of surface above domain, D

$$A(S) \;=\; \iint_D \sqrt{1 + \left(\frac{\partial z}{\partial x}\right)^2 + \left(\frac{\partial z}{\partial y}\right)^2}\; dA$$

$$A(S) \;=\; \iint_D \sqrt{1 + (z_x)^2 + (z_y)^2}\; dA$$

Surface Area -- Ex. 1a

Find the area of the part of the surface: $z = x^2 + 3y$

that lies above the triangular region in the xy-plane

with vertices at: $(0,0), (1,0), (1,1)$

Make a sketch of the domain.	

$(1, 1)$

$y = x$

$(0, 0)$

$(1, 0)$

$$A(S) = \iint_D \sqrt{1 + \left(\frac{\partial z}{\partial x}\right)^2 + \left(\frac{\partial z}{\partial y}\right)^2} \; dA$$

$$= \int_{x=0}^{x=1} \int_{y=0}^{y=x} \sqrt{1 + (2x)^2 + (3)^2} \; dy \, dx$$

$$= \int_{x=0}^{x=1} \int_{y=0}^{y=x} \sqrt{10 + 4x^2} \; dy \, dx$$

$$= \int_{x=0}^{x=1} \left[y \sqrt{10 + 4x^2} \; \right]_{y=0}^{y=x} dx$$

$$= \int_{x=0}^{x=1} \left[x \sqrt{10 + 4x^2} \; \right] dx$$

	Surface Area -- Ex. 1b

Find the area of the part of the surface: $z = x^2 + 3y$

that lies above the triangular region in the xy-plane

with vertices at: $(0,0), (1,0), (1,1)$

Previously Found	$A(S) = \int_{x=0}^{x=1} \left[x \sqrt{10 + 4x^2} \right] dx$
u-Sub	$u = 10 + 4x^2$ $du = 8x \ dx$

$A(S) = \frac{1}{8} \int_{x=0}^{x=1} (10 + 4x^2)^{\frac{1}{2}} (8 x \ dx)$

$A(S) = \frac{1}{8} \int u^{\frac{1}{2}} \ du = \frac{1}{8} \left[\frac{2}{3} \cdot u^{\frac{3}{2}} \right]$

$A(S) = \frac{1}{12} \left[(10 + 4x^2)^{\frac{3}{2}} \right]_{x=0}^{x=1}$

$A(S) = \frac{1}{12} \left[(10 + 4)^{\frac{3}{2}} - (10 + 0)^{\frac{3}{2}} \right]$

$A(S) = \frac{1}{12} \left[\sqrt{14^3} - \sqrt{10^3} \right] = \frac{7\sqrt{14} - 5\sqrt{10}}{6}$

Surface Area -- Ex. 2a

Find the area of the part of the surface: $z = x^2 + y^2$

that lies under the plane: $z = 4$

Make a sketch Domain in xy-plane is: $x^2 + y^2 = 2^2$	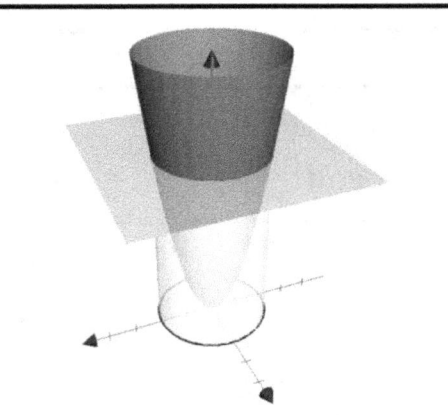

$$A(S) = \iint_D \sqrt{1 + \left(\frac{\partial z}{\partial x}\right)^2 + \left(\frac{\partial z}{\partial y}\right)^2} \ dA$$

$$A(S) = \iint_D \sqrt{1 + (2x)^2 + (2y)^2} \ dA$$

$$A(S) = \iint_D \sqrt{1 + 4(x^2 + y^2)} \ dA$$

Convert to Polar Coordinates ...

$$A(S) = \int_0^{2\pi} \int_{r=0}^{r=2} \sqrt{1 + 4r^2} \ r \ dr \ d\theta$$

Surface Area -- Ex. 2b
Find the area of the part of the surface: $z = x^2 + y^2$ that lies under the plane: $z = 4$

$$A(S) = \int_0^{2\pi} \int_{r=0}^{r=2} \sqrt{1 + 4r^2} \; r \, dr \, d\theta$$

u-Sub	$u = 1 + 4r^2$ $du = 8r \, dr$

$$A(S) = \frac{1}{8} \int_0^{2\pi} \int u^{\frac{1}{2}} \; du \, d\theta$$

$$A(S) = \frac{1}{8} \int_0^{2\pi} \left[\left(\frac{2}{3}\right) u^{\frac{3}{2}} \right] d\theta$$

$$A(S) = \frac{1}{12} \int_0^{2\pi} \left[u^{\frac{3}{2}} \right] d\theta$$

$$A(S) = \frac{1}{12} \int_0^{2\pi} \left[(1 + 4r^2)^{\frac{3}{2}} \right]_{r=0}^{r=2} d\theta$$

$$A(S) = \frac{1}{12} \int_0^{2\pi} \left[17^{\frac{3}{2}} \right] - \left[1^{\frac{3}{2}} \right] d\theta$$

$$A(S) = \frac{1}{12} \int_0^{2\pi} \left[17^{\frac{3}{2}} \right] - \left[1^{\frac{3}{2}} \right] d\theta$$

Surface Area -- Ex. 2c

Find the area of the part of the surface: $z = x^2 + y^2$

that lies under the plane: $z = 4$

$$A(S) = \frac{1}{12} \int_0^{2\pi} \left[17^{\frac{3}{2}} \right] - \left[1^{\frac{3}{2}} \right] \; d\theta$$

$$A(S) = \frac{1}{12} \int_0^{2\pi} \left(17^{\frac{3}{2}} - 1 \right) \; d\theta$$

$$A(S) = \frac{1}{12} \left[\left(17^{\frac{3}{2}} - 1 \right) \theta \right]_0^{2\pi}$$

$$A(S) = \frac{1}{12} \left[\left(17^{\frac{3}{2}} - 1 \right) 2\pi \right]$$

$$A(S) = \frac{\pi}{6} \left(17^{\frac{3}{2}} - 1 \right)$$

$$A(S) = \frac{\pi}{6} \left(17\sqrt{17} - 1 \right) \quad \approx \quad 36.18$$

Triple Integrals

Triple Integrals

- **Single** integrals are for functions of one variable.

- **Double** integrals are for functions of two variables.

- **Triple** integrals are for functions of three variables.

Triple Integral, over a box, B

$$\iiint_B f(x,y,z) \ dV$$

$$\int_r^s \int_c^d \int_a^b f(x,y,z) \ dx \, dy \, dz$$

Double Integral, over region, D	$\iint_D f(x,y) \ dx \, dy$
Single Integral, over $a \leq x \leq b$	$\int_a^b f(x) \ dx$

Triple Integrals -- Iterated

For example, consider this integral.

$$\int_0^1 \int_0^{x^2} \int_0^y f(x, y, z) \ dz \ dy \ dx$$

1.	Integrate the inner integral, first, WRT z and evaluate from: $z = 0$ to $z = y$
2.	Integrate the middle integral WRT y and evaluate from: $y = 0$ to $y = x^2$
3.	Integrate the outer integral WRT x and evaluate from: $x = 0$ to $x = 1$

Notes:
- The boundaries of the first integral (the outer integral) must be numbers, NOT functions of the other variables.
- The boundaries of the inner integrals may be functions of the other variables.
- WRT = With respect to.

Triple Integrals -- Ex. 1a

Evaluate the triple integral,

$$\iiint_B xy^2z \;\; dV$$

Where the rectangular box, B is given by:

$$B = \{\,(x,y,z)|\,0 \le x \le 1,\; 0 \le y \le 3,\,-1 \le z \le 2\,\}$$

Any order of integration is okay, but ...

The y-variable looks more complex than the other

variables so integrate it first. The x-variable is simple

and the range of x is simple, so integrate it last.

$$\int_{x=0}^{x=1}\int_{z=-1}^{z=2}\int_{y=0}^{y=3} (xy^2z)\;\; dy\;dz\;dx$$

$$\int_{x=0}^{x=1}\int_{z=-1}^{z=2} \left[\, xz\,\tfrac{1}{3}y^3\,\right]_{y=0}^{y=3}\;\; dz\;dx$$

$$\int_{x=0}^{x=1}\int_{z=-1}^{z=2} [\,xz\,9\,]\;\; dz\;dx$$

$$9\int_{x=0}^{x=1}\int_{z=-1}^{z=2} xz\;\; dz\;dx$$

Triple Integrals -- Ex. 1b

Evaluate the triple integral,

$$\iiint_B xy^2z \ dV$$

Where the rectangular box, B is given by:

$$B = \{ (x, y, z) | \ 0 \le x \le 1, \ 0 \le y \le 3, -1 \le z \le 2 \}$$

Previously found ...

$$9 \int_{x=0}^{x=1} \int_{z=-1}^{z=2} xz \ dz \ dx$$

$$9 \int_{x=0}^{x=1} \left[x \left(\frac{1}{2} \right) z^2 \right]_{z=-1}^{z=2} dx$$

$$\frac{9}{2} \int_{x=0}^{x=1} x \left[z^2 \right]_{z=-1}^{z=2} dx$$

$$\frac{9}{2} \int_{x=0}^{x=1} x \left[(2)^2 - (-1)^2 \right] dx$$

$$\frac{9}{2} \int_{x=0}^{x=1} x \left[3 \right] dx$$

$$\frac{27}{2} \int_{x=0}^{x=1} x \ dx$$

Triple Integrals -- Ex. 1c

Evaluate the triple integral,

$$\iiint_B xy^2z \ dV$$

Where the rectangular box, B is given by:

$$B = \{ (x, y, z) | \ 0 \le x \le 1, \ 0 \le y \le 3, -1 \le z \le 2 \}$$

$$\frac{27}{2} \int_{x=0}^{x=1} x \ dx \qquad \text{(Previously Found)}$$

$$\frac{27}{2} \left[\left(\frac{1}{2} \right) x^2 \right]_{x=0}^{x=1}$$

$$\frac{27}{4} \left[x^2 \right]_{x=0}^{x=1}$$

$$\frac{27}{4} \left[(1)^2 - (0)^2 \right] \quad = \quad \frac{27}{4}$$

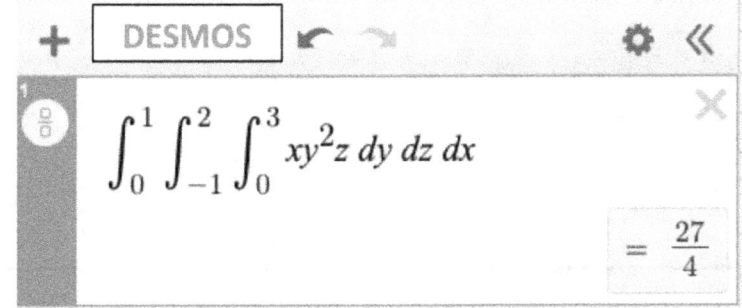

Triple Integrals -- Ex. 2a

Evaluate the triple integral: $\iiint_E z \; dV$

Where E is the solid tetrahedron
bounded by the four planes: :
$$x = 0, \quad y = 0, \quad z = 0, \quad and \quad x + y + z = 2$$

Start by making a 3D sketch. Show $z = f(x, y)$	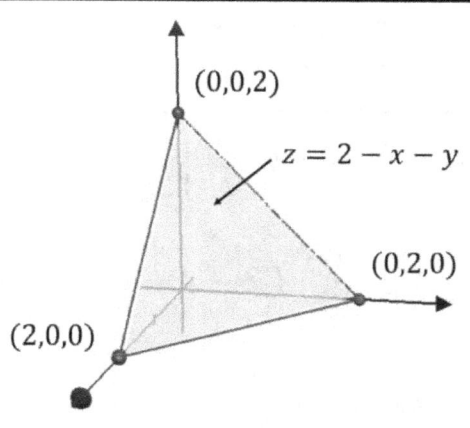
Also make a 2D sketch of the xy plane. Show $y = f(x)$	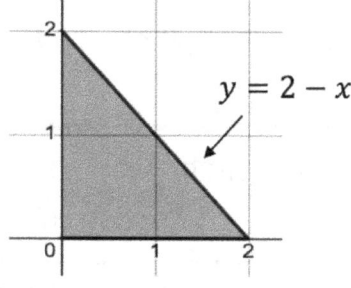

$$\iiint_E z \; dV \;=\; \int_0^2 \int_0^{2-x} \int_{z=0}^{z=2-x-y} z \; dz \, dy \, dx$$

Triple Integrals -- Ex. 2b

Evaluate the triple integral: $\iiint_E z \ dV$

Where E is the solid tetrahedron
bounded by the four planes: :
$x = 0, \quad y = 0, \quad z = 0, \quad and \quad x + y + z = 2$

$\int_0^2 \int_0^{2-x} \int_{z=0}^{z=2-x-y} z \ dz \ dy \ dx$ (Previously Found)

$\int_0^2 \int_0^{2-x} \left[\left(\frac{1}{2} \right) z^2 \right]_{z=0}^{z=2-x-y} dy \ dx$

$\frac{1}{2} \int_0^2 \int_0^{2-x} (2 - x - y)^2 \ dy \ dx$

$\frac{1}{2} \int_0^2 \left[\int_{y=0}^{y=2-x} f(x,y) \ dy \right] dx$

$\frac{1}{2} \int_{x=0}^{x=2} [\ g(x) \ ] \ dx$

The algebra can get very tedious and complex.
So, use computer software!

Triple Integrals -- Ex. 2c

Evaluate the triple integral: $\iiint_E z \; dV$

Where E is the solid tetrahedron
bounded by the four planes: :
$$x = 0, \quad y = 0, \quad z = 0, \quad and \quad x + y + z = 2$$

$\int_0^2 \int_0^{2-x} \int_{z=0}^{z=2-x-y} z \; dz \, dy \, dx$ (Previously Found)

Easily solved with Desmos ...

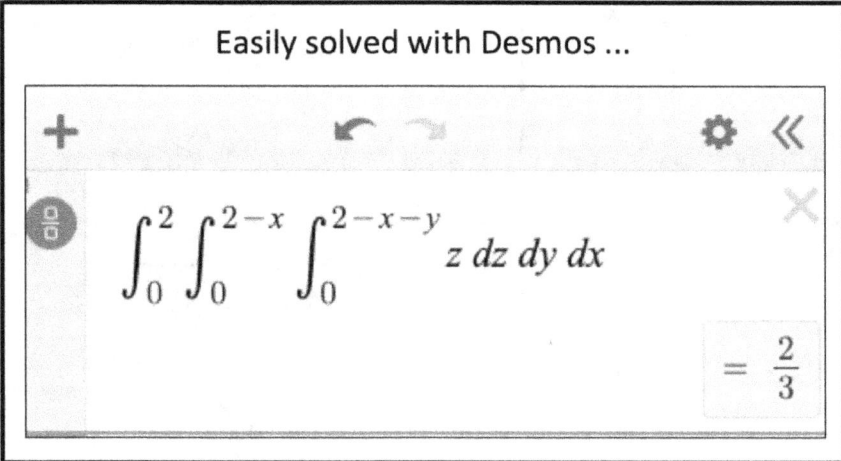

Triple Integrals -- Ex. 3a

Evaluate the triple integral

$$I = \iiint_E \sqrt{x^2 + z^2} \ dV$$

Where E is the region bounded by the

paraboloid: $y = x^2 + z^2$ and plane: $y = 4$

Start by making a 3D sketch. Here: $y = f(x, z)$	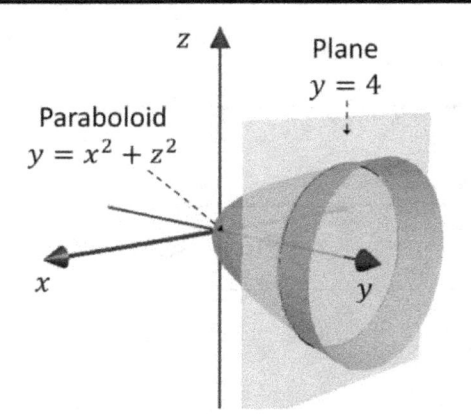
Also make a 2D sketch of the xy plane. Show $y = f(x)$	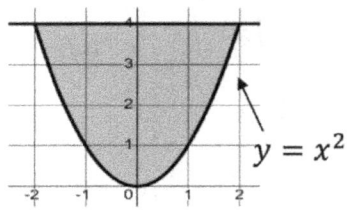

$$I = \int_{-2}^{2} \int_{x^2}^{4} \int_{-\sqrt{y - x^2}}^{\sqrt{y - x^2}} \sqrt{x^2 + z^2} \ dz \, dy \, dx$$

(Stewart, Calculus Early Transcendentals, p. 1032)

Triple Integrals -- Ex. 3b

Evaluate the triple integral

$$I = \iiint_E \sqrt{x^2 + z^2} \; dV$$

Where E is the region bounded by the

paraboloid: $y = x^2 + z^2$ \qquad and \qquad plane: $y = 4$

$$I = \int_{-2}^{2} \int_{x^2}^{4} \int_{-\sqrt{y-x^2}}^{\sqrt{y-x^2}} \sqrt{x^2 + z^2} \; dz \; dy \; dx$$

Although this expression is correct, the triple integral

is difficult to evaluate. There are several options:

- Change it to a double integral and/or convert it

  to polar coordinates.

- Use computer software (Desmos) to evaluate it.

Evaluated with Desmos ...

$+$ ↰ ↱   ⚙ ≪

    ✕

$$\int_{-2}^{2} \int_{x^2}^{4} \int_{-\sqrt{y-x^2}}^{\sqrt{y-x^2}} \sqrt{x^2 + z^2} \; dz \; dy \; dx$$

$$= 26.8082573106$$

Triple Integrals -- Ex. 4a

Use a triple integral to find the volume of the tetrahedron, T, bounded by the four planes:
$$x + 2y + z = 2, \quad x = 2y, \quad x = 0, \quad \& \quad z = 0$$

Note: This volume was previously calculated, using a double integral, in the section, Double Integrals Over General Regions, example 2.

Start by making a 3D sketch.	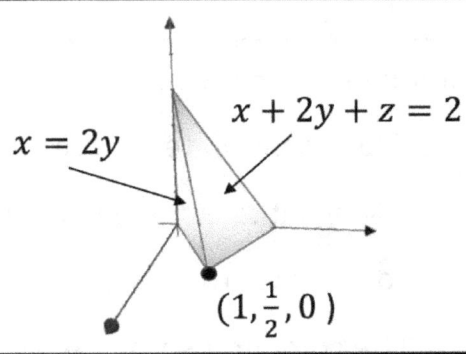 $x = 2y$ $x + 2y + z = 2$ $(1, \frac{1}{2}, 0)$
Also make a 2D sketch of the xy plane.	$y = 1 - \frac{x}{2}$ $y = \frac{x}{2}$

$$\iiint_T dV = \int_0^1 \int_{\frac{x}{2}}^{1 - \frac{x}{2}} \int_0^{2-x-2y} dz \, dy \, dx$$

(Stewart, Calculus Early Transcendentals, p. 1035)

Triple Integrals -- Ex. 4b

Use a triple integral to find the volume of the
tetrahedron, T , bounded by the four planes:
$x + 2y + z = 2, \quad x = 2y, \quad x = 0, \quad \& \quad z = 0$

Start by making a 3D sketch.	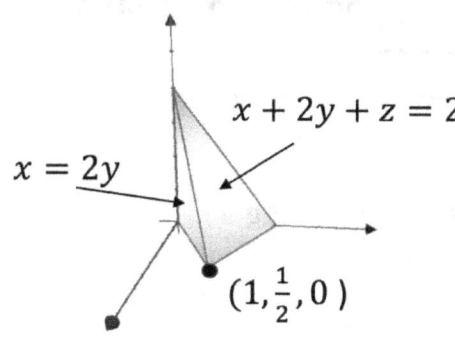
Also make a 2D sketch of the xy plane.	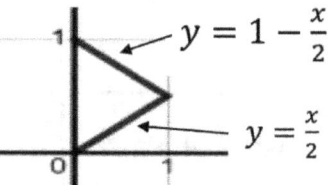

$$\iiint_T dV \;=\; \int_0^1 \int_{\frac{x}{2}}^{1-\frac{x}{2}} \int_0^{2-x-2y} dz\, dy\, dx$$

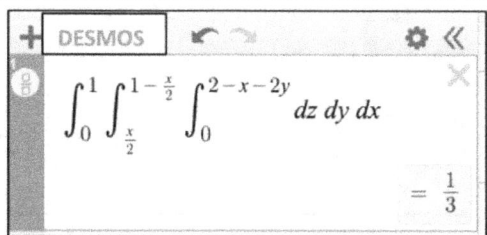

Triple Integrals
Compared to Double Integrals -- Ex. 5a

Use a **double** integral to find the volume of the
tetrahedron, T, bounded by the four planes:
$$x + 2y + z = 2, \quad x = 2y, \quad x = 0, \quad \& \quad z = 0$$
Compare to results of previous example.

Start by making a 3D sketch.	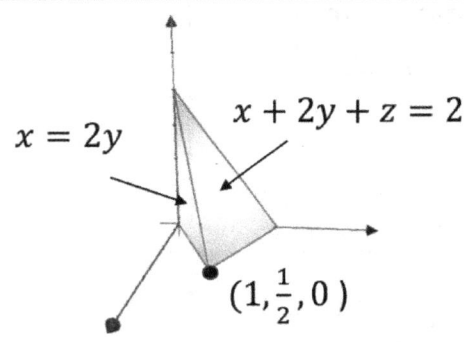
Also make a 2D sketch of the xy plane.	

$$V = \iint_D z \, dA$$

$$V = \int_0^1 \int_{\frac{1}{2}x}^{1-\frac{1}{2}x} (2 - x - 2y) \, dy \, dx$$

Triple Integrals
Compared to Double Integrals -- Ex. 5b

Use a **double** integral to find the volume of the tetrahedron, T , bounded by the four planes:
$$x + 2y + z = 2, \quad x = 2y, \quad x = 0, \quad \& \quad z = 0$$
Compare to results of previous example.

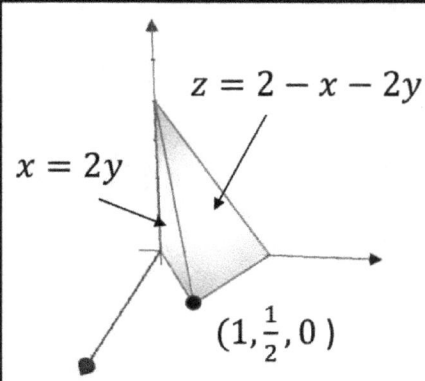

$z = 2 - x - 2y$

$x = 2y$

$(1, \tfrac{1}{2}, 0)$

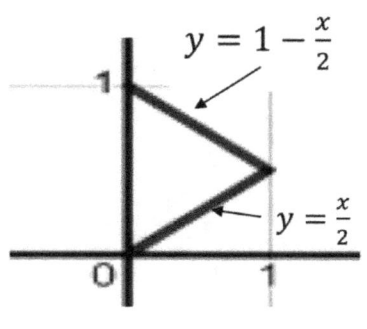

$y = 1 - \dfrac{x}{2}$

$y = \dfrac{x}{2}$

Comparison of Double and Triple Integrals using Desmos

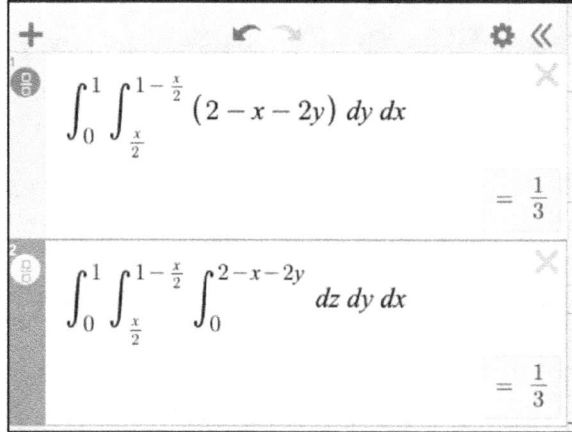

$$\int_0^1 \int_{\frac{x}{2}}^{1-\frac{x}{2}} (2 - x - 2y)\, dy\, dx$$

$$= \frac{1}{3}$$

$$\int_0^1 \int_{\frac{x}{2}}^{1-\frac{x}{2}} \int_0^{2-x-2y} dz\, dy\, dx$$

$$= \frac{1}{3}$$

Triple Integrals in Cylindrical Coordinates

Cylindrical Coordinate System

The rectangular coordinate system represents a point in 3D by the ordered triple (x, y, z)

The cylindrical coordinate system represents a point in 3D by the ordered triple (r, θ, z) Where:

$$r^2 = x^2 + y^2 \qquad \tan \theta = \frac{y}{x} \qquad z = z$$

$r = \sqrt{2^2 + 4^2} = 2\sqrt{5}$

$\theta = \tan^{-1}\left(\frac{4}{2}\right) \approx 60°$

$z = 3$

$(x, y, z) = (2, 4, 3)$

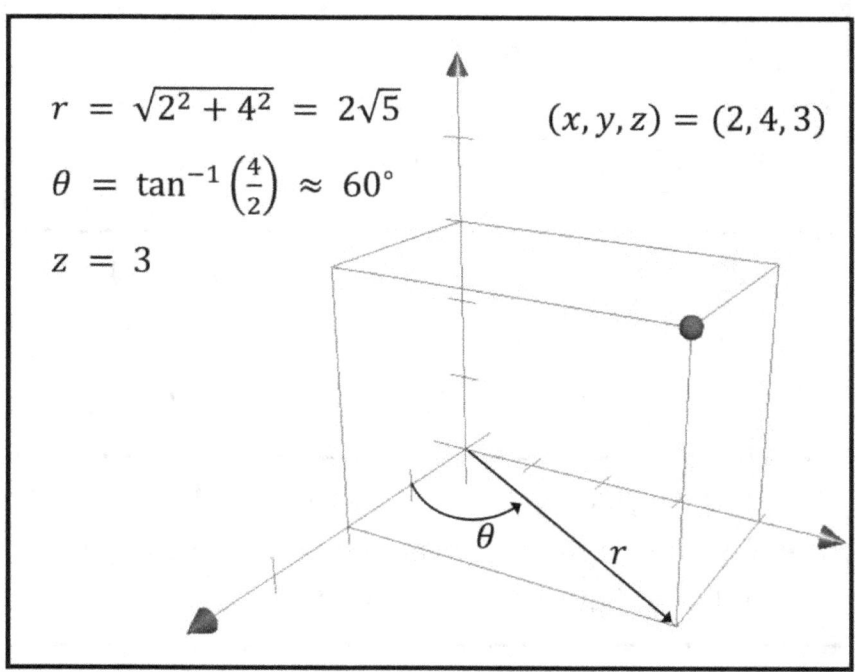

Triple Integrals In Cylindrical Coordinates

Cylindrical Coordinates are useful in 3D problems with symmetry about one axis. Setup the problem so there is symmetry about the z axis.

$$r^2 = x^2 + y^2 \qquad \tan \theta = \frac{y}{x} \qquad z = z$$

$$x = r \cos \theta \qquad y = r \sin \theta$$

$$\iiint_E f(x,y,z) \ dV \ =$$

$$= \int_{\theta_1}^{\theta_2} \int_{r_1}^{r_2} \int_{z_1}^{z_2} f(r,\theta,z) \ r \ dz \ dr \ d\theta$$

$z_1 = u_1(r,\theta)$	$z_2 = u_2(r,\theta)$
$r_1 = h_1(\theta)$	$u_2 = h_2(\theta)$
$\theta_1 = \alpha$	$\theta_2 = \beta$

Note: The <u>inside</u> integrals may have boundaries that are functions of the other variables. But, the <u>outside</u> integral must have boundaries that are fixed values.

Triple Integrals In Cylindrical Coordinates -- Ex. 1a

Find the volume that lies
- Within the cylinder: $x^2 + y^2 = 1$
- Below the plane: $z = 5$
- Above the paraboloid: $z = 1 - x^2 - y^2$

Use these equations	$r^2 = x^2 + y^2$ $\tan \theta = \dfrac{y}{x}$ $z = z$	$x = r \cos \theta$ $y = r \sin \theta$

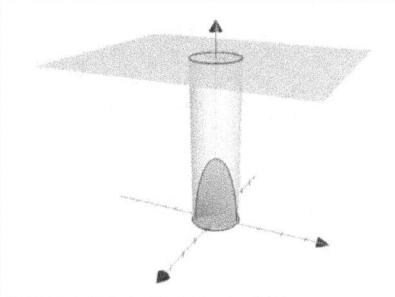

Paraboloid Surface

$z = 1 - x^2 - y^2$

$z = 1 - (x^2 + y^2)$

$z = 1 - r^2$

$V = \iiint_E f(x, y, z) \; dV =$

$= \int_0^{2\pi} \int_{r=0}^{r=1} \int_{z=1-r^2}^{z=5} (1) \; r \; dz \; dr \; d\theta$

$= \int_0^{2\pi} \int_{r=0}^{r=1} \int_{z=1-r^2}^{z=5} (r) \quad dz \; dr \; d\theta$

$= \int_0^{2\pi} \int_{r=0}^{r=1} [r \cdot z]_{z=1-r^2}^{z=5} \quad dr \; d\theta$

Triple Integrals In Cylindrical Coordinates -- Ex. 1b

Find the volume that lies

- Within the cylinder: $x^2 + y^2 = 1$
- Below the plane: $z = 5$
- Above the paraboloid: $z = 1 - x^2 - y^2$

Previously found ...

$$V = \int_0^{2\pi} \int_{r=0}^{r=1} [\, r \cdot z \,]_{z=1-r^2}^{z=5} \quad dr\, d\theta$$

$$V = \int_0^{2\pi} \int_{r=0}^{r=1} r[\, z \,]_{z=1-r^2}^{z=5} \quad dr\, d\theta$$

$$V = \int_0^{2\pi} \int_{r=0}^{r=1} r\,[\, 5 - (1 - r^2)\,] \quad dr\, d\theta$$

$$V = \int_0^{2\pi} \int_{r=0}^{r=1} r\,[\, 4 + r^2 \,] \quad dr\, d\theta$$

$$V = \int_0^{2\pi} \int_{r=0}^{r=1} [\, 4r + r^3 \,] \quad dr\, d\theta$$

$$V = \int_0^{2\pi} \left[\, 2r^2 + \left(\tfrac{1}{4}\right) r^4 \,\right]_{r=0}^{r=1} \quad d\theta$$

$$V = \int_0^{2\pi} \left[\, 2 + \tfrac{1}{4} \,\right] d\theta$$

Triple Integrals In Cylindrical Coordinates -- Ex. 1c

Find the volume that lies
- Within the cylinder: $x^2 + y^2 = 1$
- Below the plane: $z = 5$
- Above the paraboloid: $z = 1 - x^2 - y^2$

Previously found ...

$V = \int_0^{2\pi} \left[2 + \frac{1}{4} \right] d\theta$

$V = \int_0^{2\pi} \left[\frac{9}{4} \right] d\theta$

$V = \frac{9}{4} \int_0^{2\pi} [\theta]_{\theta = 0}^{\theta = 2\pi}$

$V = \frac{9}{4} [2\pi] = \frac{9\pi}{2} \approx 14.14$

Check
with
Desmos

$\int_0^{2\pi} \int_0^1 \int_{1-r^2}^5 (r)\, dz\, dr\, d\theta$

$= 14.1371669412$

Triple Integrals In Cylindrical Coordinates -- Ex. 2a

Find <u>mass</u> of the volume that lies
- Within the cylinder: $x^2 + y^2 = 1$
- Below the plane: $z = 5$
- Above the paraboloid: $z = 1 - x^2 - y^2$

The density is not constant. The density is proportional to the distance from the z-axis.

The density is: $\rho = K\sqrt{x^2 + y^2}$, $K = 30$

Use these equations	$r^2 = x^2 + y^2$ $\tan\theta = \dfrac{y}{x}$ $z = z$	$x = r\cos\theta$ $y = r\sin\theta$

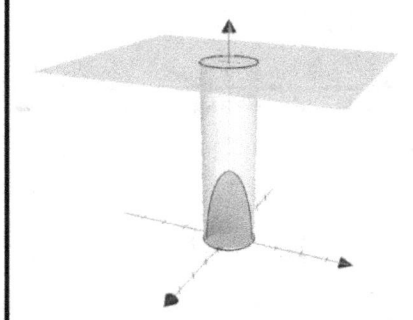

Paraboloid Surface

$z = 1 - x^2 - y^2$

$z = 1 - (x^2 + y^2)$

$z = 1 - r^2$

Mass	$\rho = 30\sqrt{x^2 + y^2}$ $\rho = 30\,r$

Triple Integrals In Cylindrical Coordinates -- Ex. 2b

Find <u>mass</u> of the volume that lies
- Within the cylinder: $x^2 + y^2 = 1$
- Below the plane: $z = 5$
- Above the paraboloid: $z = 1 - x^2 - y^2$

The density is not constant. The density is proportional to the distance from the z-axis.

The density is: $\rho = K\sqrt{x^2 + y^2}$, $K = 30$

$m = \iiint_E (\rho)\, dV$ 　　　　 Where: $\rho = 30\, r$

$m = \int_0^{2\pi} \int_{r=0}^{r=1} \int_{z=1-r^2}^{z=5} (\rho)\ r\, dz\, dr\, d\theta$

$m = \int_0^{2\pi} \int_0^1 \int_{z=1-r^2}^{z=5} (30\, r^2)\ dz\, dr\, d\theta$

$m = 30 \int_0^{2\pi} \int_0^1 r^2\, [\, z\,]_{z=1-r^2}^{z=5}\ dr\, d\theta$

$m = 30 \int_0^{2\pi} \int_0^1 r^2\, [\, 4 + r^2\,]\ dr\, d\theta$

$m = 30 \int_0^{2\pi} \int_0^1 [\, 4r^2 + r^4\,]\ dr\, d\theta$

Triple Integrals In Cylindrical Coordinates -- Ex. 2c

Find <u>mass</u> of the volume that lies
- Within the cylinder: $x^2 + y^2 = 1$
- Below the plane: $z = 5$
- Above the paraboloid: $z = 1 - x^2 - y^2$

The density is not constant. The density is proportional to the distance from the z-axis.

The density is: $\rho = K\sqrt{x^2 + y^2}$, $K = 30$

Continued ...

$$m = 30 \int_0^{2\pi} \int_0^1 [\, 4r^2 + r^4 \,] \quad dr \, d\theta$$

$$m = 30 \int_0^{2\pi} \left[\tfrac{4}{3} r^3 + \tfrac{1}{5} r^5 \right]_{r=0}^{r=1} d\theta$$

$$m = 30 \int_0^{2\pi} \left[\tfrac{20}{15} r^3 + \tfrac{3}{15} r^5 \right]_{r=0}^{r=1} d\theta$$

$$m = 2 \int_0^{2\pi} [\, 20\, r^3 + 3\, r^5 \,]_0^1 \; d\theta$$

$$m = 2 \int_0^{2\pi} [\, 23 \,] \; d\theta$$

Triple Integrals In Cylindrical Coordinates -- Ex. 2d

Find <u>mass</u> of the volume that lies
- Within the cylinder: $x^2 + y^2 = 1$
- Below the plane: $z = 5$
- Above the paraboloid: $z = 1 - x^2 - y^2$

The density is not constant. The density is proportional to the distance from the z-axis.

The density is: $\rho = K\sqrt{x^2 + y^2}$, $K = 30$

Continued ...

$$m = 2 \int_0^{2\pi} [\,23\,] \ d\theta$$

$$m = 46 \,[\,\theta\,]_0^{2\pi}$$

$$m = 46\,(\,2\pi\,) \ = \ 92\pi \ \approx \ 289.03$$

Check
with
Desmos

$+$ ↰ ↱   ⚙

$$\int_0^{2\pi} \int_0^1 \int_{1-r^2}^5 \left(30\,r^2\right) dz\ dr\ d\theta$$

$$= \ 289.026524$$

Triple Integrals In Cylindrical Coordinates -- Ex. 3a

Find <u>mass</u> of the volume that lies
- Within the cylinder: $x^2 + y^2 = 1$
- Below the plane: $z = 5$
- Above the paraboloid: $z = 1 - x^2 - y^2$

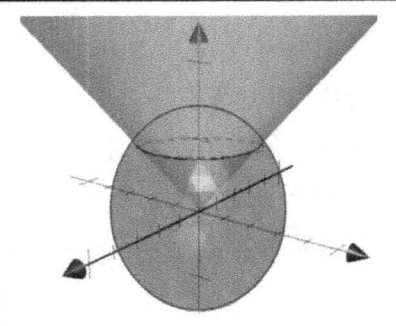

For cylindrical

coordinates, use:

$$x^2 + y^2 = r^2$$

Intersection → $0 \le r \le 1$	$x^2 + y^2 = 2 - x^2 - y^2$ $2(x^2 + y^2) = 2$ $x^2 + y^2 = 1$
Sphere:	$(x^2 + y^2) + z^2 = 2$ $z = \sqrt{2 - r^2}$
Cone:	$z = \sqrt{x^2 + y^2} = r$
Range for z	$r \le z \le \sqrt{2 - r^2}$

Triple Integrals In Cylindrical Coordinates -- Ex. 3b

Find <u>mass</u> of the volume that lies
- Within the cylinder: $x^2 + y^2 = 1$
- Below the plane: $z = 5$
- Above the paraboloid: $z = 1 - x^2 - y^2$

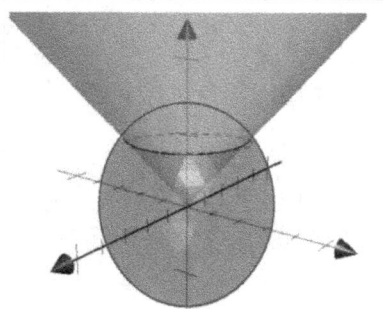

Previously found:

$$0 \leq r \leq 1$$

$$r \leq z \leq \sqrt{2 - r^2}$$

$$V = \int_0^{2\pi} \int_0^1 \int_r^{\sqrt{2-r^2}} r \, dz \, dr \, d\theta$$

$$V = \int_0^{2\pi} \int_0^1 [\, r z \,]_r^{\sqrt{2-r^2}} \, dr \, d\theta$$

$$V = \int_0^{2\pi} \int_0^1 [\, r\sqrt{2 - r^2} - r^2 \,] \, dr \, d\theta$$

$$u = 2 - r^2 \, , \quad du = -2r \, dr$$

$$V = \int_0^{2\pi} \left[\left(-\frac{1}{2} \right) \left(\frac{2}{3} \right) (2 - r^2)^{\frac{3}{2}} - \frac{1}{3} r^3 \right]_0^1 \, d\theta$$

Triple Integrals In Cylindrical Coordinates -- Ex. 3c

Find <u>mass</u> of the volume that lies
- Within the cylinder: $\quad x^2 + y^2 = 1$
- Below the plane: $\quad z = 5$
- Above the paraboloid: $\quad z = 1 - x^2 - y^2$

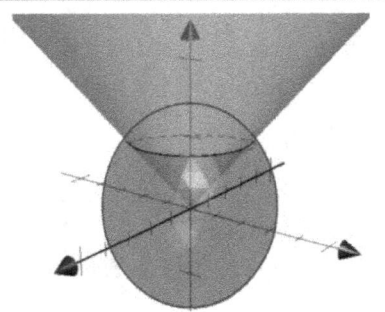

Previously found:

$$0 \le r \le 1$$

$$r \le z \le \sqrt{2 - r^2}$$

$$V = \int_0^{2\pi} \left[\left(-\frac{1}{2} \right) \left(\frac{2}{3} \right) (2 - r^2)^{\frac{3}{2}} - \frac{1}{3} r^3 \right]_0^1 \, d\theta$$

$$V = \int_0^{2\pi} \left[\left(\left(-\frac{1}{3} \right) (1) - \frac{1}{3} \right) - \left(\left(-\frac{1}{3} \right) 2^{\frac{3}{2}} \right) \right] \, d\theta$$

$$V = \int_0^{2\pi} \left[-\frac{2}{3} + \frac{2\sqrt{2}}{3} \right] \, d\theta$$

$$V = (2\pi) \left(\frac{2}{3} \right) [-1 + \sqrt{2}] \quad = \quad \frac{4\pi}{3} (\sqrt{2} - 1)$$

Triple Integrals In Cylindrical Coordinates -- Ex. 3d

Find <u>mass</u> of the volume that lies
- Within the cylinder: $x^2 + y^2 = 1$
- Below the plane: $z = 5$
- Above the paraboloid: $z = 1 - x^2 - y^2$

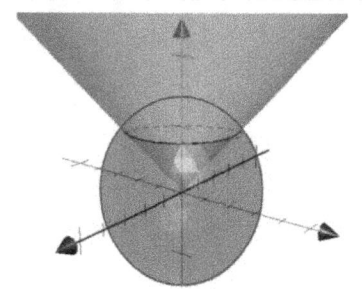

Previously found:

$$0 \leq r \leq 1$$

$$r \leq z \leq \sqrt{2 - r^2}$$

Summary ...

$$V = \int_0^{2\pi} \int_0^1 \int_r^{\sqrt{2-r^2}} r \, dz \, dr \, d\theta$$

$$V = \frac{4\pi}{3} \left(\sqrt{2} - 1 \right) \approx 1.735$$

Check with Desmos

$$\int_0^{2\pi} \int_0^1 \int_r^{\sqrt{2-r^2}} r \, dz \, dr \, d\theta$$

$$= 1.73505371276$$

Triple Integrals in Spherical Coordinates

Spherical Coordinate System

The **spherical coordinate system** represents a point in 3D by the ordered triple (ρ, θ, ϕ)

Where:

$$\rho = \sqrt{x^2 + y^2 + z^2}$$

$$x = \rho \cdot \sin\phi \cdot \cos\theta$$

$$y = \rho \cdot \sin\phi \cdot \sin\theta$$

$$z = \rho \cdot \cos\phi \quad , \quad 0 \le \phi \le \pi$$

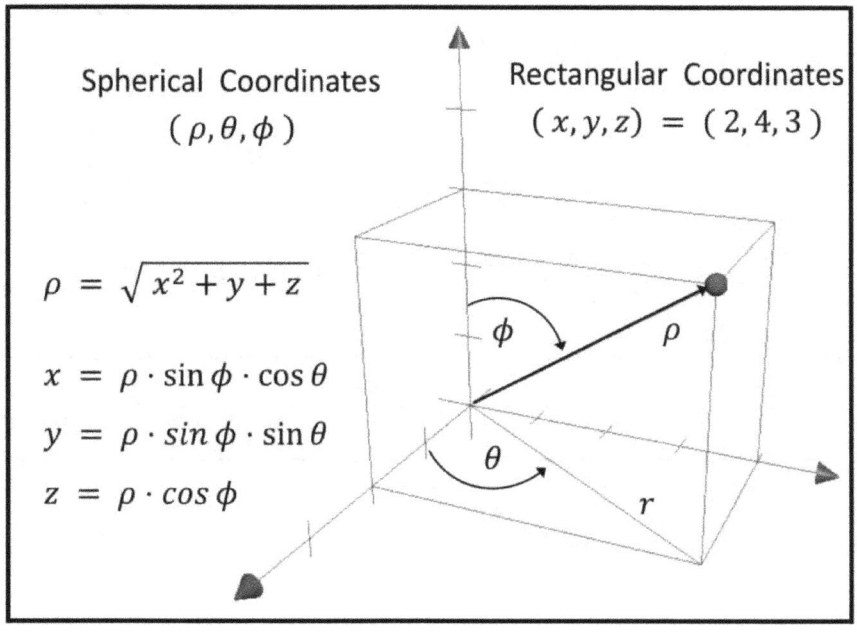

Spherical Coordinates
(ρ, θ, ϕ)

Rectangular Coordinates
$(x, y, z) = (2, 4, 3)$

$\rho = \sqrt{x^2 + y + z}$

$x = \rho \cdot \sin\phi \cdot \cos\theta$

$y = \rho \cdot \sin\phi \cdot \sin\theta$

$z = \rho \cdot \cos\phi$

Triple Integrals in Spherical Coordinates

Spherical Coordinates are useful in 3D problems with symmetry about one point. Setup the problem so there is symmetry about the origin, point $(0,0,0)$.

$$\rho = \sqrt{x^2 + y^2 + z^2}$$

$$x = \rho \cdot \sin\phi \cdot \cos\theta$$

$$y = \rho \cdot \sin\phi \cdot \sin\theta$$

$$z = \rho \cdot \cos\phi \, , \quad 0 \le \phi \le \pi$$

$$\iiint_E f(x,y,z) \; dV \; =$$

$$\int_{\phi_1}^{\phi_2} \int_{\theta_1}^{\theta_2} \int_{\rho_1}^{\rho_2} f(\rho,\theta,\phi) \; (\rho^2 \sin\phi \; d\rho \; d\theta \; d\phi)$$

$$\int_{\phi_1}^{\phi_2} \int_{\theta_1}^{\theta_2} \int_{\rho_1}^{\rho_2} f(\rho,\theta,\phi) \; \rho^2 \sin\phi \; d\rho \; d\theta \; d\phi$$

Note: $\quad dV \; = \; (\rho^2 \sin\phi \; d\rho \; d\theta \; d\phi)$

Triple Integrals in Spherical Coordinates -- Ex. 1a

Use spherical coordinates to find
the volume of a sphere with radius = 3.

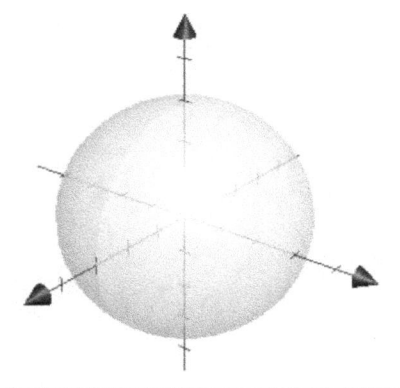

$$0 \leq \phi \leq \pi$$

$$0 \leq \theta \leq 2\pi$$

$$0 \leq \rho \leq 3$$

$$V = \iiint_E f(x, y, z) \ dV =$$

$$V = \int_0^\pi \int_0^{2\pi} \int_0^3 \ (1) \ \ \rho^2 \sin \phi \ d\rho \ d\theta \ d\phi$$

$$V = \int_0^\pi \int_0^{2\pi} \int_0^3 \ (\rho^2 \sin \phi) \ \ d\rho \ d\theta \ d\phi$$

$$V = \int_0^\pi \int_0^{2\pi} \ \sin \phi \ \left[\frac{1}{3} \rho^3 \right]_0^3 \ d\theta \ d\phi$$

$$V = \int_0^\pi \int_0^{2\pi} \ \sin \phi \ [9] \ d\theta \ d\phi$$

Triple Integrals in Spherical Coordinates -- Ex. 1b

Use spherical coordinates to find
the volume of a sphere with radius = 3.

Previously found ...

$$V = \int_0^\pi \int_0^{2\pi} \sin\phi \; [9] \; d\theta \, d\phi$$

$$V = 9 \int_0^\pi \int_0^{2\pi} (\sin\phi) \, d\theta \, d\phi$$

$$V = 9 \int_0^\pi \sin\phi \; [\theta]_0^{2\pi} \, d\phi$$

$$V = 9 \int_0^\pi \sin\phi \; [2\pi] \, d\phi$$

$$V = 18\pi \int_0^\pi \sin\phi \; d\phi$$

$$V = 18\pi \; [-\cos\phi]_0^\pi$$

$$V = 18\pi \; [(-\cos\pi) - (-\cos 0)]$$

$$V = 18\pi \; [2] \quad = \quad 36\pi$$

Check	$V = \frac{4}{3} \pi r^3 = \frac{4}{3} \pi (3)^3 = 36\pi$

Triple Integrals in Spherical Coordinates -- Ex. 2a

Use spherical coordinates to find
the volume the solid that lies

- Above the cone: $\quad z = \sqrt{x^2 + y^2}\quad$ and
- Below the sphere: $\quad z = x^2 + y^2 + z^2$

Find where cone and sphere intersect.

$$z = \sqrt{x^2 + y^2}$$

$$\rho \cos\phi = \sqrt{(\rho \sin\phi \cos\theta)^2 + (\rho \sin\phi \sin\theta)^2}$$

$$\rho \cos\phi = \sqrt{(\rho \sin\phi)^2 (\cos^2\theta + \sin^2\theta)}$$

$$\rho \cos\phi = \sqrt{(\rho \sin\phi)^2 (1)}$$

$$\rho \cos\phi = \rho \sin\phi$$

$$\cos\phi = \sin\phi \qquad \rightarrow \qquad \phi = \pi/4$$

Cone and sphere intersect at: $\quad \phi = \pi/4$

(Stewart, Calculus Early Transcendentals, p. 1048)

Triple Integrals in Spherical Coordinates -- Ex. 2b

Use spherical coordinates to find
the volume the solid that lies

- Above the cone: $z = \sqrt{x^2 + y^2}$ and
- Below the sphere: $z = x^2 + y^2 + z^2$

Previously found ...

Cone and sphere intersect at: $\phi = \pi/4$

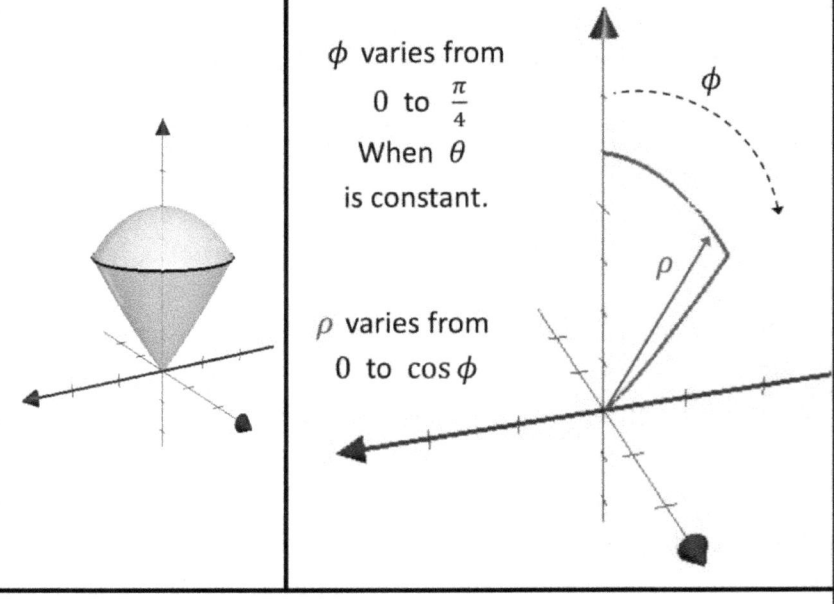

ϕ varies from

    0 to $\dfrac{\pi}{4}$

When θ
is constant.

ρ varies from

    0 to $\cos\phi$

$$V = \int_0^{2\pi} \int_0^{\frac{\pi}{4}} \int_0^{\cos\phi} (1)\ \rho^2 \sin\phi\ d\rho\, d\phi\, d\theta$$

Triple Integrals in Spherical Coordinates -- Ex. 2c

Use spherical coordinates to find
the volume the solid that lies

- Above the cone: $z = \sqrt{x^2 + y^2}$ and
- Below the sphere: $z = x^2 + y^2 + z^2$

Previously found ...

$$V = \int_0^{2\pi} \int_0^{\frac{\pi}{4}} \int_0^{\cos\phi} (1) \ \rho^2 \sin\phi \ d\rho \ d\phi \ d\theta$$

$$V = \left[\int_0^{2\pi} (1) \ d\theta \right] \cdot \int_0^{\frac{\pi}{4}} \int_0^{\cos\phi} (\rho^2 \sin\phi) \ d\rho \ d\phi$$

$$V = [\, 2\pi \,] \cdot \int_0^{\frac{\pi}{4}} \sin\phi \ \left[\frac{1}{3} \rho^3 \right]_0^{\cos\phi} d\phi$$

$$V = \left[\frac{2\pi}{3} \right] \cdot \int_0^{\frac{\pi}{4}} \sin\phi \cdot \cos^3\phi \ d\phi$$

U-Substitution	$u = \cos\phi \ , \quad du = -\sin\phi$

$$V = \left[-\frac{2\pi}{3} \right] \cdot \int u^3 \ du = \left[-\frac{2\pi}{3} \right] \cdot \left[\frac{1}{4} u^4 \right]$$

$$V = \left[-\frac{\pi}{6} \right] \cdot [\, \cos^4\phi \,]_0^{\frac{\pi}{4}}$$

Triple Integrals in Spherical Coordinates -- Ex. 2d

Use spherical coordinates to find
the volume the solid that lies

- Above the cone: $z = \sqrt{x^2 + y^2}$ and
- Below the sphere: $z = x^2 + y^2 + z^2$

Previously found ...

$$V = \left[-\frac{\pi}{6} \right] \cdot \left[\cos^4 \phi \right]_0^{\frac{\pi}{4}}$$

$$V = \left[-\frac{\pi}{6} \right] \cdot \left[\left(\frac{\sqrt{2}}{2}\right)^4 - (1)^4 \right]$$

$$V = \left[-\frac{\pi}{6} \right] \cdot \left[-\frac{12}{16} \right] = \frac{\pi}{8} \approx .393$$

Check answer with Desmos ...

$$\int_0^{2\pi} \int_0^{\frac{\pi}{4}} \int_0^{\cos\phi} \left(\rho^2 \cdot \sin\phi \right) d\rho \, d\phi \, d\theta$$

$$= 0.392699081699$$

Triple Integrals in Spherical Coordinates -- Ex. 3a	

Use spherical coordinates to evaluate

$$\iiint_B (x^2 + y^2 + z^2)\ dV$$

Where B is the ball, centered at the origin,
with radius $= R$

Spherical Coordinates	$z = \rho \cos\phi$ $x = r\cos\theta = \rho\sin\phi\cos\theta$ $y = r\sin\theta = \rho\sin\phi\sin\theta$ $r = \rho\sin\phi$ $\rho^2 = x^2 + y^2 + z^2$
Ranges	$0 \le \rho \le R$ $0 \le \theta \le 2\pi$ $0 \le \phi \le \pi$

$\iiint_E f(x,y,z)\ dV =$

$\quad = \iiint_B f(x,y,z)\ \ \rho^2 \sin\phi\ \ d\rho\ d\theta\ d\phi$

$\quad = \int_0^\pi \int_0^{2\pi} \int_0^R (\rho^2)\ \ \rho^2 \sin\phi\ \ d\rho\ d\theta\ d\phi$

Triple Integrals in Spherical Coordinates -- Ex. 3b

Use spherical coordinates to evaluate

$$\iiint_B (x^2 + y^2 + z^2)\ dV$$

Where B is the ball, centered at the origin, with radius $= R$

$\iiint_E f(x,y,z)\ dV\ =$

$= \int_0^\pi \int_0^{2\pi} \int_0^R (\rho^2)\ \rho^2 \sin\phi\ d\rho\ d\theta\ d\phi$

$= \int_0^\pi \int_0^{2\pi} \int_0^R (\rho^4 \sin\phi)\ d\rho\ d\theta\ d\phi$

$= \int_0^\pi \int_0^{2\pi} \left[\frac{1}{5}\rho^5 \sin\phi\right]_0^R\ d\theta\ d\phi$

$= \frac{R^5}{5} \int_0^\pi \int_0^{2\pi} \sin\phi\ d\theta\ d\phi$

$= \frac{R^5}{5} (2\pi) \int_0^\pi \sin\phi\ d\phi$

$= \frac{R^5}{5} (2\pi)[-\cos\phi]_0^\pi$

$= -\frac{R^5}{5} (2\pi)[-1-1]\ =\ \frac{4\pi}{5} R^5$

Triple Integrals in Spherical Coordinates -- Ex. 4a

Use spherical coordinates to evaluate

$$\iiint_E \left(x^2 \right) dV$$

Where E is the solid hemisphere:

$$x^2 + y^2 + z^2 \leq 25, \qquad x \geq 0$$

Spherical Coordinates	$z = \rho \cos \phi$ $x = r \cos \theta = \rho \sin \phi \cos \theta$ $y = r \sin \theta = \rho \sin \phi \sin \theta$ $r = \rho \sin \phi$ $\rho^2 = x^2 + y^2 + z^2$
Ranges	$0 \leq \rho \leq R$ $0 \leq \theta \leq 2\pi$ $0 \leq \phi \leq \dfrac{\pi}{2}$ $\leftarrow$ hemisphere

$$\iiint_E f(x,y,z) \; dV$$

$$= \iiint_E (x^2) \; \rho^2 \sin \phi \;\; d\rho \; d\theta \; d\phi$$

$$= \int_0^{\frac{\pi}{2}} \int_0^{2\pi} \int_0^5 (\rho \sin \phi \cos \theta)^2 \, \rho^2 \sin \phi \; d\rho \; d\theta \; d\phi$$

Triple Integrals in Spherical Coordinates -- Ex. 4b
Use spherical coordinates to evaluate $$\iiint_E \left(x^2 \right) dV$$ Where E is the solid hemisphere: $$x^2 + y^2 + z^2 \le 25, \qquad x \ge 0$$

$\iiint_E \left(x^2 \right) dV =$

$= \iiint_E \left(x^2 \right) \rho^2 \sin\phi \ \ d\rho \ d\theta \ d\phi$

$= \int_0^{\frac{\pi}{2}} \int_0^{2\pi} \int_0^5 \left(\rho \sin\phi \cos\theta \right)^2 \rho^2 \sin\phi \ d\rho \ d\theta \ d\phi$

$= \int_0^{\frac{\pi}{2}} \int_0^{2\pi} \int_0^5 \left(\rho^4 \sin^3\phi \cos^2\theta \right) \quad d\rho \ d\theta \ d\phi$

$= \int_0^{\frac{\pi}{2}} \int_0^{2\pi} \left[\frac{1}{5} \rho^5 \sin^3\phi \cos^2\theta \right]_0^5 \ d\theta \ d\phi$

$= \int_0^{\frac{\pi}{2}} \int_0^{2\pi} \left[625 \ \sin^3\phi \cos^2\theta \right] \ d\theta \ d\phi$

$= 625 \int_0^{\frac{\pi}{2}} \sin^3\phi \ d\phi \ \int_0^{2\pi} \cos^2\theta \ d\theta$

$= 625 \left(I_1 \right) \left(I_2 \right)$

Triple Integrals in Spherical Coordinates -- Ex. 4c

Use spherical coordinates to evaluate
$$\iiint_E (x^2)\, dV$$
Where E is the solid hemisphere:
$$x^2 + y^2 + z^2 \leq 25, \qquad x \geq 0$$

$\iiint_E (x^2)\, dV =$

$= 625 \int_0^{\frac{\pi}{2}} \sin^3 \phi \; d\phi \;\; \int_0^{2\pi} \cos^2 \theta \; d\theta$

$= 625\, (I_1)\,(I_2)$

$I_1 = \int_0^{\frac{\pi}{2}} \sin^3 \phi \; d\phi$

$I_1 = \int_0^{\frac{\pi}{2}} \sin^2 \phi \, \sin \phi \; d\phi$

$I_1 = \int_0^{\frac{\pi}{2}} (1 - \cos^2 \phi)\, \sin \phi \; d\phi$

$I_1 = \left[-\cos \phi - \frac{1}{3}\cos^3 \phi \right]_0^{\frac{\pi}{2}}$

$I_1 = \left[(0 + 0) - \left(-1 + \frac{1}{3}\right) \right] \quad = \quad \frac{2}{3}$

Change of Variable in Multiple Integrals

Change of Variables

Variables can be changed or transformed into new variables to make equations easier to work with.

For example, suppose we have a complicated function of variables (x, y) and we want to transform the function into a simpler function of variables (u, v). This is like using U-sub. on two variables.

If T is a one-to-one transformation, then it has an inverse transformation, T^{-1}
as shown below.

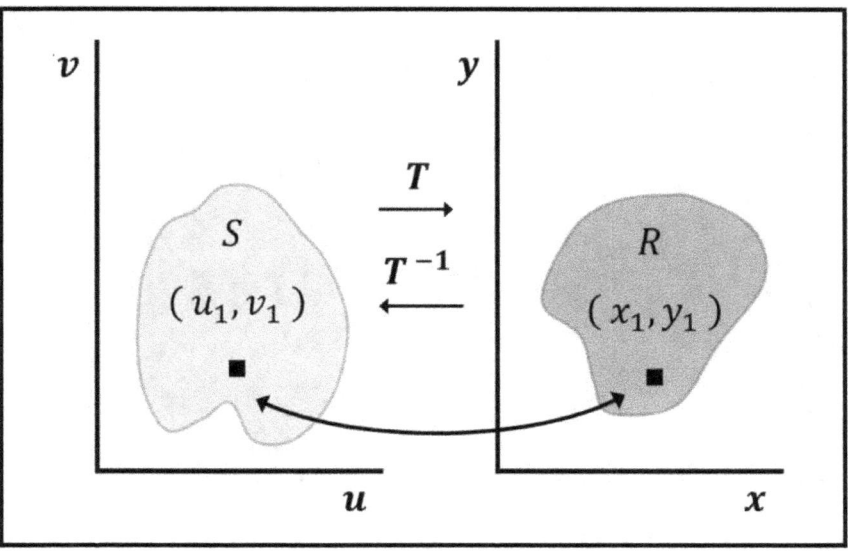

Jacobian of the Transformation − 2 x 2
$$x = g(u, v) \quad \text{and} \quad y = h(u, v)$$

The Jacobian is used to transform variables for

integration.   We will do this later.

First, learn how to calculate the Jacobian.

The Jacobian transformation T, given by

$$x = g(u, v) \quad \text{and} \quad y = h(u, v) \text{ is:}$$

$$J = \frac{\delta(x, y)}{\delta(u, v)}$$

$$J = \begin{vmatrix} \dfrac{\delta x}{\delta u} & \dfrac{\delta x}{\delta v} \\[2mm] \dfrac{\delta y}{\delta u} & \dfrac{\delta y}{\delta v} \end{vmatrix} = \begin{vmatrix} x_u & x_v \\[2mm] y_u & y_v \end{vmatrix}$$

$$J = \frac{\delta x}{\delta u} \cdot \frac{\delta y}{\delta v} \ - \ \frac{\delta x}{\delta v} \cdot \frac{\delta y}{\delta u}$$

$$J = x_u \cdot y_v \ - \ x_v \cdot y_u$$

Jacobian of the Transformation – 3 x 3

$$x = g(u, v, w) \ , \ \ y = h(u, v, w) \ \& \ \ z = h(u, v, w)$$

The Jacobian transformation T, given by

$$x = x(u, v, w) \ , \ \ y = y(u, v, w) \ \text{ and}$$

$$z = z(u, v, w) \ \text{ is:}$$

$$J = \frac{\delta(x, y, z)}{\delta(u, v, w)}$$

$$J = \begin{vmatrix} x_u & x_v & x_w \\ y_u & y_v & y_w \\ z_u & z_v & z_w \end{vmatrix}$$

$$J = x_u (y_v \cdot z_w - y_w \cdot z_v)$$

$$- x_v (y_u \cdot z_w - y_w \cdot z_u)$$

$$+ x_w (y_u \cdot z_v - y_v \cdot z_u)$$

Change of Variables -- Double Integrals

$$\iint_R f(x, y)\, dA \;\; = \;\; \iint_S f(u, v)\; |J|\; du\, dv$$

Where: $J \;\; = \;\; \dfrac{\delta\,(x,y)}{\delta\,(u,v)}$

Change of Variables -- Triple Integrals

$$\iiint_R f(x, y, z)\, dV \;\; = \;\; \iiint_S f(u, v, w)\; |J|\; du\, dv\, dw$$

Where: $J \;\; = \;\; \dfrac{\delta\,(x,y,z)}{\delta\,(u,v,w)}$

Change of Variables -- Double Integrals -- Ex. 1a

Evaluate the integral: $I = \iint_R y \, dA$

Where R is the region bounded by parabolas:

$y^2 = 4 - 4x$, $y^2 = 4 + 4x$, $y \geq 0$

Given the change of variables in S

$x = u^2 - v^2$ and $y = 2uv$

Where S is the square: $[0,1] \times [0,1]$

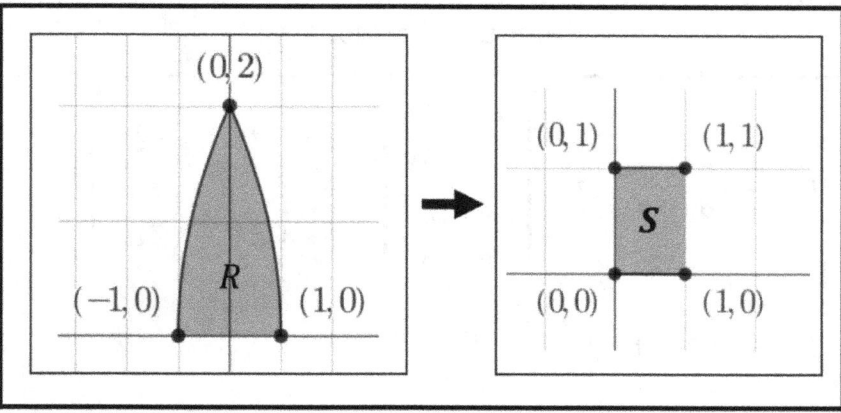

(Stewart, Calculus Early Transcendentals, p. 1057)

Change of Variables -- Double Integrals -- Ex. 1b

Evaluate the integral: $\quad I = \iint_R y \, dA$

Where R is the region bounded by parabolas:

$y^2 = 4 - 4x \quad , \quad y^2 = 4 + 4x \quad , \quad y \geq 0$

Given the change of variables in S

$\quad x = u^2 - v^2 \quad$ and $\quad y = 2uv$

- Where S is the square: $[0,1] \times [0,1]$

Compute the Jacobian

$$J \; = \; \frac{\delta\,(x,y)}{\delta\,(u,v)} \; = \; \begin{vmatrix} \dfrac{\delta x}{\delta u} & \dfrac{\delta x}{\delta v} \\[2mm] \dfrac{\delta y}{\delta u} & \dfrac{\delta y}{\delta v} \end{vmatrix} \; = \; \begin{vmatrix} 2u & -2v \\[2mm] 2v & 2u \end{vmatrix}$$

$$J \; = \; 4u^2 + 4v^2 \; > \; 0$$

Change of Variables -- Double Integrals -- Ex. 1c

Evaluate the integral: $I = \iint_R y \, dA$

Where R is the region bounded by parabolas:

$y^2 = 4 - 4x$, $y^2 = 4 + 4x$, $y \geq 0$

Given the change of variables in S

$x = u^2 - v^2$ and $y = 2uv$

Where S is the square: $[0,1] \times [0,1]$

Previously found the Jacobian:

$J = 4u^2 + 4v^2 > 0$

$I = \iint_R y \, dA = \iint_S 2uv \, |J| \, dA$

$I = \int_{v=0}^{v=1} \int_{u=0}^{u=1} (2uv)(4u^2 + 4v^2) \, du \, dv$

$I = 8 \int_{v=0}^{v=1} \int_{u=0}^{u=1} (u^3 v + uv^3) \, du \, dv$

$I = 8 \int_{v=0}^{v=1} \left[\frac{1}{4} u^4 v + \frac{1}{2} u^2 v^3 \right]_0^1 \, dv$

$I = \int_0^1 (2v + 4v^3) \, dv = [v^2 - v^4]_0^1 = 2$

Change of Variables -- Double Integrals -- Ex. 1d

Evaluate the integral: $I = \iint_R y \, dA$

Where R is the region bounded by parabolas:

$y^2 = 4 - 4x$, $y^2 = 4 + 4x$, $y \geq 0$

Given the change of variables in S

   $x = u^2 - v^2$ and $y = 2uv$

Where S is the square: $[0,1] \times [0,1]$

Extra: Desmos graph of region, R , with $y \geq 0$
And the Desmos code used to create it.

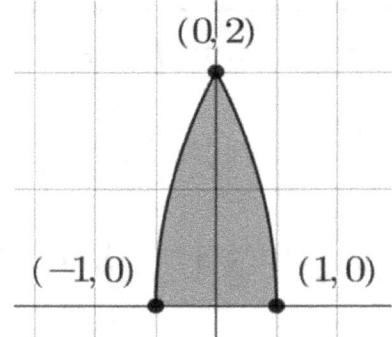

$(0, 2)$

$(-1, 0)$ $(1, 0)$

$$\frac{\left(y^2 - 4\right)}{4} \leq x \leq \frac{\left(y^2 - 4\right)}{-4} \quad \{y \geq 0\}$$

Change of Variables -- Double Integrals -- Ex. 2a

Evaluate the integral: $I = \iint_R e^{\frac{(x+y)}{(x-y)}} dA$

Where R is a trapezoid region with vertices:

$$(x, y) = (1,0), (2,0), (0, -2), (0, -1)$$

Identify the change of variables.	Let: $u = (x + y)$ $v = (x - y)$

Convert (x, y) points to (u, v) points		

(x, y)	$(x + y, x - y)$
$(1, 0)$	$(1, 1)$
$(2, 0)$	$(2, 2)$
$(0, -2)$	$(-2, 2)$
$(0, -1)$	$(-1, 1)$

New S Region	Has vertices at points $(u, v) =$ $(1,1), (2,2), (-2,2), (-1,1)$

(Stewart, Calculus Early Transcendentals, p. 1057)

Change of Variables -- Double Integrals -- Ex. 2b

Evaluate the integral: $I = \iint_R e^{\frac{(x+y)}{(x-y)}} \, dA$

Where R is a trapezoid region with vertices:

$(x, y) = (1,0), (2,0), (0, -2), (0, -1)$

Previously Found	S region vertices at $(u, v) =$ $(1,1), (2,2), (-2,2), (-1,1)$

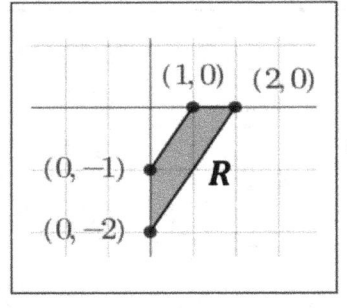
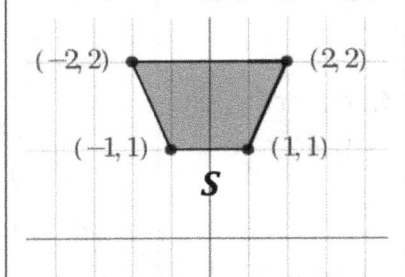

Previously found	$u = (x + y)$ $v = (x - y)$

Find x & y In terms of u & v	$x = u - y$ $y = x - v$	$x = \frac{1}{2}(u + v)$ $y = \frac{1}{2}(u - v)$

Change of Variables -- Double Integrals -- Ex. 2c		
Evaluate the integral: $I = \iint_R e^{\frac{(x+y)}{(x-y)}} dA$ Where R is a trapezoid region with vertices: $(x, y) = (1,0), (2,0), (0,-2), (0,-1)$		

Previously Found	$u = (x + y)$ $\qquad$ $v = (x - y)$	
	$x = \frac{1}{2}(u + v) \qquad y = \frac{1}{2}(u - v)$	
Find partial derivs.	$x_u = \frac{1}{2}$	$x_v = \frac{1}{2}$
	$y_u = \frac{1}{2}$	$y_v = -\frac{1}{2}$
Find the Jacobian	$J = \begin{vmatrix} x_u & x_v \\ y_u & y_v \end{vmatrix} = \begin{vmatrix} \frac{1}{2} & \frac{1}{2} \\ \frac{1}{2} & -\frac{1}{2} \end{vmatrix}$	
	$J = -\frac{1}{4} - \frac{1}{4} = -\frac{1}{2}$	

Change of Variables -- Double Integrals -- Ex. 2d

Evaluate the integral: $I = \iint_R e^{\frac{(x+y)}{(x-y)}} \, dA$

Where R is a trapezoid region with vertices:

$(x, y) = (1,0), (2,0), (0, -2), (0, -1)$

Previously Found	$$J = -\frac{1}{2}$$
Previously Found	
Boundaries for S	$1 \leq v \leq 2$ & $-v \leq u \leq v$

Use the jacobian and the boundaries to setup the integral in region S Continued ...

Change of Variables -- Double Integrals -- Ex. 2e

Evaluate the integral: $I = \iint_R e^{\frac{(x+y)}{(x-y)}} dA$

Where R is a trapezoid region with vertices:

$(x, y) = (1,0), (2,0), (0,-2), (0,-1)$

Previously Found	$J = -\dfrac{1}{2}$
Boundaries for S	$1 \leq v \leq 2 \quad \& \quad -v \leq u \leq v$

Transform integral in R to an integral in S
Then evaluate it.

$I = \iint_R e^{\frac{(x+y)}{(x-y)}} dA$

Jacobian of the Transformation

$I = \iint_S e^{\frac{u}{v}} \left| \dfrac{\delta\,(x,y)}{\delta\,(u,v)} \right| du\,dv$

$I = \int_1^2 \int_{-v}^{v} e^{\frac{u}{v}} \left| -\dfrac{1}{2} \right| du\,dv$

$I = \int_1^2 \int_{-v}^{v} e^{\frac{u}{v}} \left(\dfrac{1}{2}\right) du\,dv$

Change of Variables -- Double Integrals -- Ex. 2f

Evaluate the integral: $I = \iint_R e^{\frac{(x+y)}{(x-y)}} dA$

Where R is a trapezoid region with vertices:

$(x, y) = (1,0), (2,0), (0,-2), (0,-1)$

Previously Found	$I = \int_1^2 \int_{-v}^{v} e^{\frac{u}{v}} \left(\frac{1}{2}\right) du\, dv$

$I = \frac{1}{2} \int_1^2 \int_{-v}^{v} e^{\frac{u}{v}} \, du\, dv$

$I = \frac{1}{2} \int_1^2 \int_{-v}^{v} e^u v^{-1} \, du\, dv$

$I = \frac{1}{2} \int_1^2 (v) \int_{-v}^{v} e^{u \left(\frac{1}{v}\right)} \left(\frac{1}{v}\right) du\, dv$

$I = \frac{1}{2} \int_1^2 (v) \left[e^{u \left(\frac{1}{v}\right)} \right]_{u=-v}^{u=v} dv$

$I = \frac{1}{2} \int_1^2 (v) \left[e - e^{-1} \right] dv$

$I = \frac{1}{2} (e - e^{-1}) \int_1^2 (v) \, dv$

Change of Variables -- Double Integrals -- Ex. 2g

Evaluate the integral: $I = \iint_R e^{\frac{(x+y)}{(x-y)}} \, dA$

Where R is a trapezoid region with vertices:

$$(x, y) = (1,0), (2,0), (0, -2), (0, -1)$$

Previously Found	$I = \frac{1}{2} (e - e^{-1}) \int_1^2 (v) \, dv$
$I = \frac{1}{2} \left(e - \frac{1}{e} \right) \int_1^2 (v) \, dv$	
$I = \frac{1}{2} \left(\frac{e^2 - 1}{e} \right) \left[\frac{1}{2} v^2 \right]_{v=1}^{v=2}$	
$I = \frac{1}{4} \left(\frac{e^2 - 1}{e} \right) \left[v^2 \right]_{v=1}^{v=2}$	
$I = \frac{1}{4} \left(\frac{e^2 - 1}{e} \right) [4 - 1]$	
$I = \frac{3}{4} \left(\frac{e^2 - 1}{e} \right) \approx 1.76$	
Desmos Check	$\int_1^2 \int_{-v}^{v} e^{\frac{u}{v}} \left(\frac{1}{2} \right) du \, dv$ $= 1.7628017904$

Change of Variables -- Double Integrals -- Ex. 2h

Evaluate the integral: $I = \iint_R e^{\frac{(x+y)}{(x-y)}} \, dA$

Using a calculator such as Desmos, we could evaluate the integral, in R if we break it into two parts. No change of variable needed.

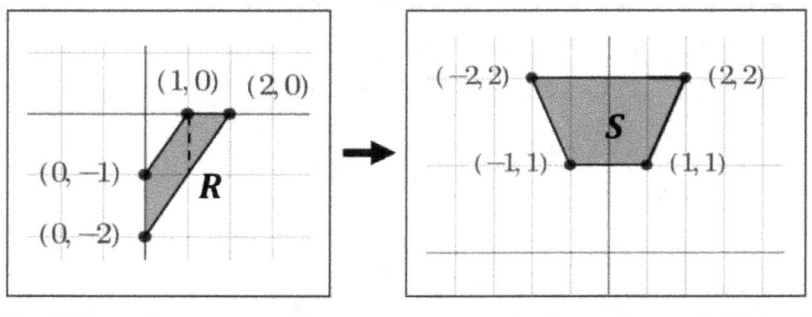

S $\displaystyle\int_1^2 \int_{-v}^{v} e^{\left(\frac{u}{v}\right)} \left(\frac{1}{2}\right) du \, dv$

$= 1.762801790\cdot$

R $\displaystyle\int_0^1 \int_{x-2}^{x-1} e^{\frac{(x+y)}{(x-y)}} \, dy \, dx + \int_1^2 \int_{x-2}^{0} e^{\frac{(x+y)}{(x-y)}} \, dy \, dx$

$= 1.762801790\cdot$

Change of Variables -- Double Integrals -- Ex. 3a

Find the volume of the solid lying below the surface: $f(x, y) = xy$ and above region R.

Where R is the region with vertices: $(x, y) =$

$(1,0), (0,1), (1,2), (2,1)$

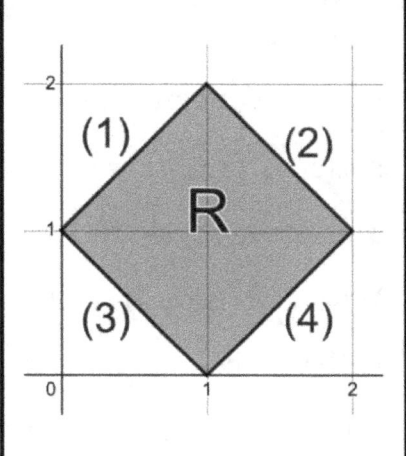

(1)	(2)
$y = 1 + x$ $y - x = 1$	$y = 3 - x$ $y + x = 3$

(3)	(4)
$y = 1 - x$ $y + x = 1$	$y = -1 + x$ $y - x = -1$

Let: $u = y - x$ $\rightarrow$ $-1 \leq u \leq 1$

Let: $v = y + x$ $\rightarrow$ $1 \leq v \leq 3$

Continued ...

Change of Variables -- Double Integrals -- Ex. 3b

Find the volume of the

solid lying below the

surface: $f(x,y) = xy$

and above region R .

Where R is the region

with vertices: $(x,y) =$

$(1,0), (0,1), (1,2), (2,1)$

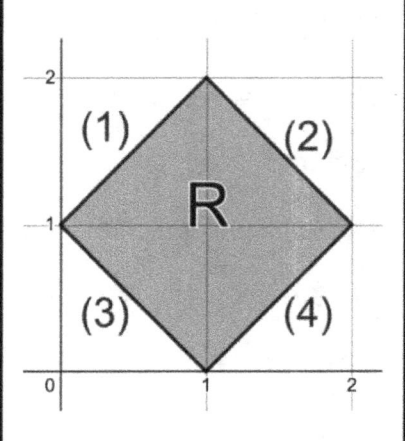

Identify the change of Variables ...

$x = y - u$	$y = v - x$
$x = (v - x) - u$	$y = v - (y - u)$
$x = \dfrac{v - u}{2}$	$y = v - y + u$
	$2y = v + u$
	$y = \dfrac{v + u}{2}$

$$f(x,y) \;=\; xy \;=\; \left(\frac{v-u}{2}\right)\left(\frac{v+u}{2}\right) \;=\; \frac{v^2 - u^2}{4}$$

Change of Variables -- Double Integrals -- Ex. 3c

Find the volume of the solid lying below the surface: $f(x,y) = xy$ and above region R.

Where R is the region with vertices: $(x,y) =$

$(1,0), (0,1), (1,2), (2,1)$

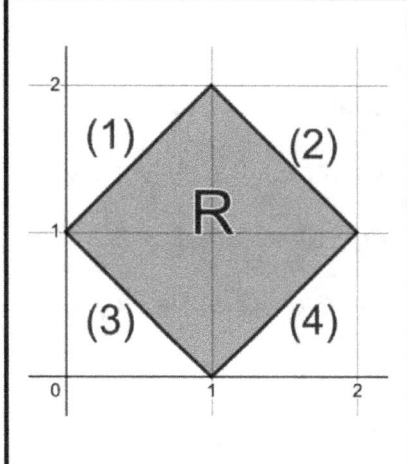

Previously Found	$x = \dfrac{v-u}{2}$ and $y = \dfrac{v+u}{2}$

$$J = \begin{vmatrix} x_u & x_v \\ y_u & y_v \end{vmatrix} = \begin{vmatrix} -\dfrac{1}{2} & \dfrac{1}{2} \\ \dfrac{1}{2} & \dfrac{1}{2} \end{vmatrix} = -\dfrac{1}{4} - \dfrac{1}{4} = -\dfrac{1}{2}$$

$$V = \iint f(x,y) \; dA$$

$$V = \iint f(u,v) \; |J| \; du \; dv$$

$$V = \int_1^3 \int_{-1}^1 \left(\frac{v^2 - u^2}{4}\right) \left|-\frac{1}{2}\right| \; du \; dv$$

Change of Variables -- Double Integrals -- Ex. 3d

Find the volume of the

solid lying below the

surface: $f(x,y) = xy$

and above region R .

Where R is the region

with vertices: $(x,y) =$

$(1,0), (0,1), (1,2), (2,1)$

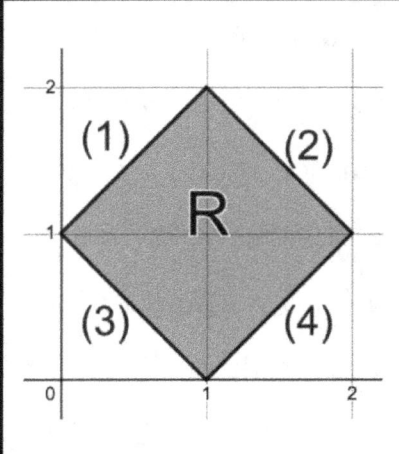

$$V = \int_1^3 \int_{-1}^1 \left(\frac{v^2 - u^2}{4} \right) \left| -\frac{1}{2} \right| \, du \, dv$$

$$V = \frac{1}{8} \int_1^3 \left[uv^2 - \frac{1}{3}u^3 \right]_{-1}^1 \, dv$$

$$V = \frac{1}{8} \int_1^3 \left[\left(v^2 - \frac{1}{3} \right) - \left(-v^2 + \frac{1}{3} \right) \right] \, dv$$

$$V = \frac{1}{8} \int_1^3 \left[2v^2 - \frac{2}{3} \right] \, dv$$

$$V = \frac{1}{8} \left[\frac{2}{3}v^3 - \frac{2}{3}v \right]_1^3 = \frac{1}{12} \left[v^3 - v \right]_1^3 = 2$$

Vector Calculus

Vector Fields & Gradient Fields

Vector Fields
A Vector Field is a function that assigns vectors to points.

Vector Fields on $\mathbb{R}^2$
A Vector Field on $\mathbb{R}^2$ is a function F that assigns to each point (x, y) in D a two-dimensional vector $F(x, y)$. Where D is a set (plane field) in $\mathbb{R}^2$.

Vector Fields on $\mathbb{R}^3$
A Vector Field on $\mathbb{R}^3$ is a function F that assigns to each point (x, y) in E a three-dimensional vector $F(x, y, z)$. Where E is a sub-set of $\mathbb{R}^3$.

Gradient Fields

Gradient Fields are vector fields.

$$\nabla f(x,y) \;=\; f_x(x,y)\,\boldsymbol{i} \;+\; f_y(x,y)\,\boldsymbol{j}$$

$$\nabla f(x,y) \;=\; \langle\, f_x, f_y \,\rangle$$

3D Gradient Fields

3D Gradient Fields are 3D vector fields.

$$\nabla f(x,y,z) \;=\; f_x\,\boldsymbol{i} \;+\; f_y\,\boldsymbol{j} \;+\; f_z\,\boldsymbol{k}$$

$$\nabla f(x,y,z) \;=\; \langle\, f_x, f_y, f_z \,\rangle.$$

Vector Fields -- Some Tools

Below are examples of two tools that may be used to graph vector fields. It is not a complete list. Using vector field graphing tools is a good way for students to easily visualize vector fields.

Both examples show the same $\mathbb{R}^2$ simple vector field function: $F(x,y) = x\,\mathbf{i} + y\,\mathbf{j}$

Google "GeoGebra" Runs best on Chrome Browser. Easy interface.	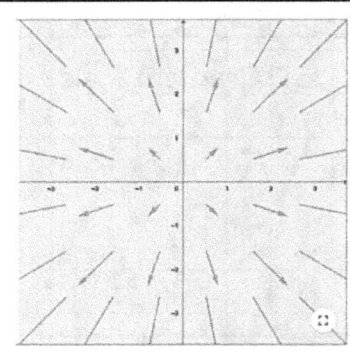
Apple "Grapher" Included with Apple computers. 2D or 3D	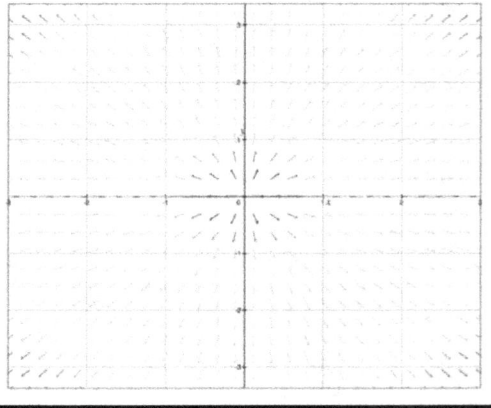

Gradient Fields -- Ex. 1a

Given: $f(x,y) = xy - x^3$

Find and plot the gradient vector field.

Use a vector graphing tool of your choice.

Gradient vector field	$\nabla f = \langle\ f_x,\ f_y\ \rangle$ $\nabla f = \langle\ y - 3x^2,\ x\ \rangle$
Google GeoGebra	
Apple Grapher	

Gradient Fields -- Ex. 1b   (A closer look)

Google GeoGebra	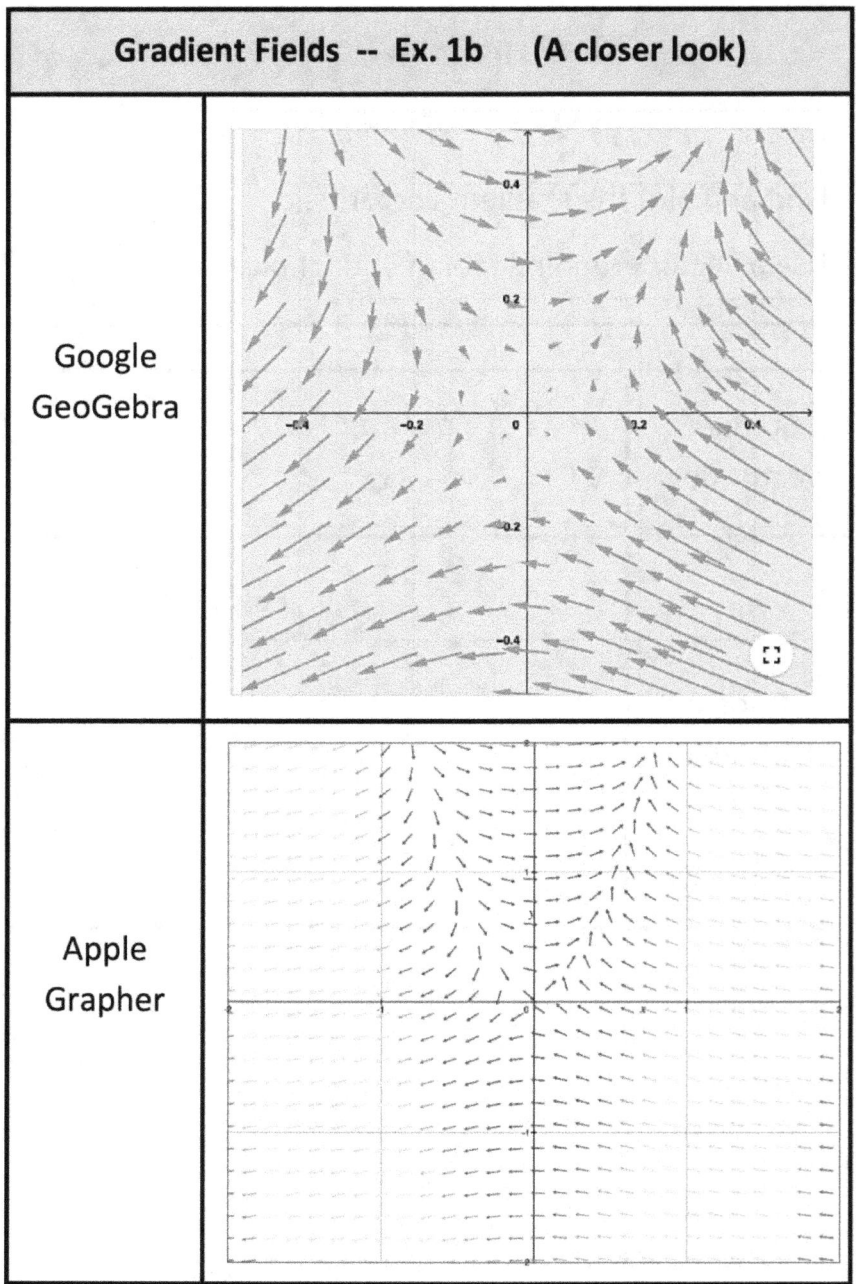
Apple Grapher	

Gradient Fields -- Ex. 2

Find the gradient vector field of f

Given: $f(x,y) = (x+y)e^{x-y}$

Gradient vector field	$\nabla f = \langle f_x, f_y \rangle$
Rewrite Equation	$f(x,y) = x e^{x-y} + y e^{x-y}$

$f_x = [1] e^{x-y} + x[e^{x-y}] + y e^{x-y}$

$f_x = e^{x-y}(1 + x + y)$

$f_y = x e^{x-y}(-1) + [1] e^{x-y} + y[e^x - y)(-1)]$

$f_y = -x e^{x-y} + e^{x-y} - y e^{x-y}$

$f_y = e^{x-y}(1 - x - y)$

$\nabla f = \langle f_x, f_y \rangle$

$\nabla f = \langle e^{x-y}(1 + x + y), e^{x-y}(1 - x - y) \rangle$

Gradient Fields -- Ex. 3

Find the gradient vector field of f

Given: $f(x, y, z) = \| \langle x, y, z \rangle \|$

Gradient vector field	$\nabla f = \langle f_x, f_y, f_z \rangle$
Rewrite Equation	$f(x, y, z) = \sqrt{x^2 + y^2 + z^2}$

$f_x = \frac{1}{2}(x^2 + y^2 + z^2)^{-\frac{1}{2}}(2x) = \dfrac{x}{\sqrt{x^2 + y^2 + z^2}}$

$f_y = \frac{1}{2}(x^2 + y^2 + z^2)^{-\frac{1}{2}}(2y) = \dfrac{y}{\sqrt{x^2 + y^2 + z^2}}$

$f_z = \frac{1}{2}(x^2 + y^2 + z^2)^{-\frac{1}{2}}(2z) = \dfrac{z}{\sqrt{x^2 + y^2 + z^2}}$

$\nabla f = \langle f_x, f_y, f_z \rangle$

$\nabla f = \dfrac{\langle x, y, z \rangle}{\sqrt{x^2 + y^2 + z^2}}$

Gradient Fields -- Ex. 4a

Given: The temp. in a metal ball is inversely proportional to the distance from the center (origin). The temperature at point $(1, 2, 2)$ is 110°.

Find: Find the Temperature rate of change at point $(1, 2, 2)$ in the direction towards point $(4, 1, 3)$.

Temp. is inversely proportional to d.	$T = k\dfrac{1}{d}$
	$T = k\ \dfrac{1}{\sqrt{x^2 + y^2 + z^2}}$
	$110 = k\dfrac{1}{\sqrt{1^2 + 2^2 + 2^2}} \quad \rightarrow \quad k = 330$
	$T = 330\,(x^2 + y^2 + z^2)^{-\frac{1}{2}}$
Vector from $(1, 2, 2)$ To $(4, 1, 3)$	$v = \langle \Delta x, \Delta y, \Delta z \rangle$
	$v = \langle 4 - 1, 1 - 2, 3 - 2 \rangle$
	$v = \langle 3, -1, 1 \rangle$
	$u = \text{unit vector} = \dfrac{v}{|v|}$
	$u = \dfrac{\langle 3,\ -1,\ 1 \rangle}{\sqrt{3^2 + 1^2 + 1^2}} = \dfrac{\langle 3,\ -1,\ 1 \rangle}{\sqrt{11}}$

Gradient Fields -- Ex. 4b

Given: The temp. in a metal ball is inversely proportional to the distance from the center (origin). The temperature at point $(1, 2, 2)$ is 110°.

Find: Find the Temperature rate of change at point $(1, 2, 2)$ in the direction towards point $(4, 1, 3)$.

Previously Found	$T = 330 \left(x^2 + y^2 + z^2 \right)^{-\frac{1}{2}}$
	$u = \dfrac{\langle 3, -1, 1 \rangle}{\sqrt{3^2 + 1^2 + 1^2}} = \dfrac{\langle 3, -1, 1 \rangle}{\sqrt{11}}$
Gradient	$\nabla T = \langle \dfrac{\delta T}{\delta x}, \dfrac{\delta T}{\delta y}, \dfrac{\delta T}{\delta z} \rangle$
Directional Derivative	$D_u T = \nabla T \cdot u$
$\dfrac{\delta T}{\delta x} = 330 \left(-\dfrac{1}{2} \right) \left(x^2 + y^2 + z^2 \right)^{-\frac{3}{2}} (2x)$	
$\dfrac{\delta T}{\delta y} = 330 \left(-\dfrac{1}{2} \right) \left(x^2 + y^2 + z^2 \right)^{-\frac{3}{2}} (2y)$	
$\dfrac{\delta T}{\delta z} = 330 \left(-\dfrac{1}{2} \right) \left(x^2 + y^2 + z^2 \right)^{-\frac{3}{2}} (2z)$	

Gradient Fields -- Ex. 4c

Given: The temp. in a metal ball is inversely proportional to the distance from the center (origin). The temperature at point $(1, 2, 2)$ is 110°.

Find: Find the Temperature rate of change at point $(1, 2, 2)$ in the direction towards point $(4, 1, 3)$.

$$\nabla T(x, y, z) \;=\; \langle \frac{\delta T}{\delta x}, \frac{\delta T}{\delta y}, \frac{\delta T}{\delta z} \rangle \qquad \text{Temp. ROC}$$

$$\nabla T(x, y, z) \;=\; \frac{-330}{(x^2 + y^2 + z^2)^{\frac{3}{2}}} \langle x, y, z \rangle$$

$$\nabla T(1,2,2) \;=\; \frac{-330}{(9)^{\frac{3}{2}}} \langle 1, 2, 2 \rangle \;=\; -\frac{110}{9} \langle 1, 2, 2 \rangle$$

$$D_u T \;=\; \nabla T \cdot u$$

$$D_u T \;=\; \left[-\frac{110}{9} \langle 1, 2, 2 \rangle \right] \cdot \left[\frac{1}{\sqrt{11}} \langle 3, -1, 1 \rangle \right]$$

$$D_u T \;=\; \left(-\frac{110}{9} \right) \left(\frac{1}{\sqrt{11}} \right) (3 - 2 + 2) \;\approx\; -11.06$$

Line Integrals

Line Integrals

For a line integral, we integrate over a curve, instead of integrating over an interval $[a, b]$.

The curve is defined by parametric equations:

$$x = x(t) \qquad y = y(t) \qquad a \leq t \leq b$$

Recall, the length of a curve is given by:

$$L = \int_a^b \sqrt{\left(\frac{dx}{dt}\right)^2 + \left(\frac{dy}{dt}\right)^2}$$

Line Integral of a function, over curve, C

$$L = \int_C f(x, y)\ ds$$

$$L = \int_a^b f(x(t), y(t)) \sqrt{\left(\frac{dx}{dt}\right)^2 + \left(\frac{dy}{dt}\right)^2}\ dt$$

$$L = \int_a^b f(x, y) \sqrt{(x')^2 + (y')^2}\ dt$$

Where:

The curve is defined by parametric equations:

$$x = x(t) \qquad y = y(t) \qquad a \leq t \leq b$$

Line Integrals -- Visual Example

Think of a line integral as the area of a fence over a curved base. For this example, the height is a constant 2 feet and the curve is a circle: $x^2 + y^2 = 9$

Parameterize The curve	$x = r \cos t = 3 \cos t$ $y = r \sin t = 3 \sin t$ $0 \le t \le 2\pi$

Area of fence $= \int_C f(x,y)\, ds$

$$A = \int_a^b f(x(t),\, y(t)) \sqrt{\left(\frac{dx}{dt}\right)^2 + \left(\frac{dy}{dt}\right)^2}\; dt$$

$$A = \int_0^{2\pi} (2) \sqrt{(3 \sin t)^2 + (3 \cos t)^2}\; dt$$

$$A = \int_0^{2\pi} (2) \sqrt{9(\sin^2 t + \cos^2 t)}\; dt$$

$$A = \int_0^{2\pi} 6\; dt = 6\,[\,t\,]_0^{2\pi} = 6\,[\,2\pi\,] = 12\pi$$

Check	Area = Circumference * height Area $= (2\pi r) \cdot h = (6\pi) \cdot 2 = 12\pi$

Line Integrals -- Ex. 1a

Evaluate the line integral: $\int_C (3 + xy^2)\, ds$

Where:

C is the upper half of a circle: $x^2 + y^2 = 4$

Sketch the curve	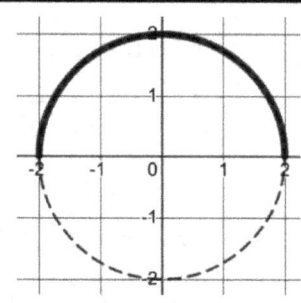
Parameterize the curve	$x = 2\cos(t)\quad,\quad y = 2\sin(t)$ $0 \le t \le \pi$

$$\int_C = \int_0^\pi f(x,y)\,\sqrt{(x')^2 + (y')^2}\ dt$$

$$= \int_0^\pi (3 + (2\cos t)(2\sin t)^2\,\sqrt{(x')^2 + (y')^2}\ dt$$

$$= \int_0^\pi (3 + 8\cos t\,\sin^2 t)\,\sqrt{(x')^2 + (y')^2}\ dt$$

$$= \int_0^\pi (3 + 8\cos t\,\sin^2 t)$$
$$\sqrt{(-2\sin t)^2 + (2\cos t)^2}\ dt$$

Line Integrals -- Ex. 1b

Evaluate the line integral: $\int_C (3 + xy^2)\, ds$

Where:

C is the upper half of a circle: $x^2 + y^2 = 4$

$\int_C = \int_0^\pi (3 + 8\cos t\, \sin^2 t)$

$$\sqrt{(-2\sin t)^2 + (2\cos t)^2}\quad dt$$

$\int_C = \int_0^\pi (3 + 8\cos t\, \sin^2 t)$

$$2\sqrt{\sin^2 t + \cos^2 t}\quad dt$$

$\int_C = \int_0^\pi (3 + 8\cos t\, \sin^2 t) \cdot 2\sqrt{1}\quad dt$

$\int_C = 2\int_0^\pi (3 + 8\cos t\, \sin^2 t)\quad dt$

$\int_C = 6\int_0^\pi 1\, dt\; +\; 16\int_0^\pi \cos t\, \sin^2 t\quad dt$

$\int_C = 6[\, t\,]_0^\pi\; +\; 16\int_0^\pi \cos t\, \sin^2 t\quad dt$

$\int_C = 6\pi\; +\; 16\int u^2\, du\; ;\qquad u = \sin t$

$\int_C = 6\pi\; +\; 16\left[\dfrac{u^3}{3}\right]$

Line Integrals -- Ex. 1c

Evaluate the line integral: $\int_C (3 + xy^2)\, ds$

Where:

C is the upper half of a circle: $x^2 + y^2 = 4$

$\int_C = 6\pi + 16\left[\dfrac{u^3}{3}\right]$; $u = \sin t$

$\int_C = 6\pi + \dfrac{16}{3}\left[\sin^3 t\right]_0^\pi$

$\int_C = 6\pi + \dfrac{16}{3}[0]$

$\int_C = 6\pi \approx 18.85$

Desmos Check ...

$$\int_0^\pi \left(3 + 8(\cos t)\sin^2 t\right) \cdot \sqrt{(2\cdot \sin t)^2 + (2\cdot \cos t)^2}\ dt$$

$$= 18.8495559211$$

Line Integrals -- Ex. 2a

Evaluate the line integral: $I = \int_C (x \cos z)\, ds$

Where: C is a circular helix: $0 \le t \le 2\pi$

With: $x = \cos t,$ $y = \sin t,$ $z = t$

Given curve	$x = \cos(t),\ \ y = \sin(t),\ \ z = t$ $0 \le t \le 2\pi$

$$I = \int_0^{2\pi} f(x,y,z)\, \sqrt{(x')^2 + (y')^2 + (z')^2}\ \ dt$$

$$I = \int_0^{2\pi} (x \cos z)\, \sqrt{(x')^2 + (y')^2 + (z')^2}\ \ dt$$

$$I = \int_0^{2\pi} (\cos t \cos t)\, \cdot$$
$$\sqrt{(-\sin t)^2 + (\cos t)^2 + (1)^2}\ \ dt$$

$$I = \int_0^{2\pi} (\cos^2 t)\cdot \sqrt{1+1}\ \ dt$$

$$I = \sqrt{2} \int_0^{2\pi} \left(\tfrac{1}{2}(1 + \cos 2t)\right) dt$$

$$I = \frac{\sqrt{2}}{2} \int_0^{2\pi} (1 + \cos 2t)\ d t$$

Line Integrals -- Ex. 2b

Evaluate the line integral: $I = \int_C (x \cos z)\, ds$

Where: C is a circular helix: $0 \le t \le 2\pi$

With: $x = \cos t,\quad y = \sin t,\quad z = t$

Given curve	$x = \cos(t),\quad y = \sin(t),\quad z = t$ $0 \le t \le 2\pi$

$I = \dfrac{\sqrt{2}}{2} \int_0^{2\pi} (1 + \cos 2t)\; dt$

$I = \dfrac{\sqrt{2}}{2} \left[t + \dfrac{1}{2}\sin 2t \right]_0^{2\pi}$

$I = \dfrac{\sqrt{2}}{2} \left[(2\pi - 0) - (0 - 0) \right]$

$I = \dfrac{\sqrt{2}}{2} \left[2\pi \right]$

$I = \pi\sqrt{2} \qquad \approx \qquad 4.44$

Desmos Check	$\sqrt{2} \cdot \displaystyle\int_0^{2\pi} \cos^2(t)\; dt$
	$= 4.44288293816$

Line Integrals -- Ex. 3

Evaluate the line integral: $I = \int_C f(x, y, z)\, ds$

Where: $f(x, y, z) = x + y + z$

And $\quad c(t) = (\sin t, \cos t, t)$, $\quad 0 \le t \le 2\pi$

$f(t) = \sin t + \cos t + t$

$c'(t) = (\cos t, -\sin t, 1)$

$|c'(t)| = \sqrt{\cos^2 t + \sin^2 t + 1} = \sqrt{2}$

$I = \int_C f \cdot ds = \int_C f(t) \cdot |c'(t)|\ dt$

$I = \int_0^{2\pi} (\sin t + \cos t + t) \cdot (\sqrt{2})\ dt$

$I = \sqrt{2} \int_0^{2\pi} (\sin t + \cos t + t)\ dt$

$I = \sqrt{2} \left[-\cos t + \sin t + \dfrac{t^2}{2} \right]_0^{2\pi}$

$I = \sqrt{2} \left[\left(-1 + 0 + \dfrac{4\pi}{2} \right) - (-1 + 0 + 0) \right]$

$I = \sqrt{2} [2\pi^2] = 2\pi^2 \sqrt{2} \approx 27.9$

<u>Line Integrals of Vector Fields</u>

Line Integrals of Vector Fields

$\int_C F \cdot dr \; = \;$ Line integral of F along curve C

$\int_C F \cdot dr \; = \; \int_a^b F(r) \cdot r' \;\; dt$

Where:

$\quad F = f(x, y, z) \; = \;$ Continuous Vector Field

$\quad r = f(t) \; = \;$ Vector function

Line Integrals of Vector Fields -- Ex. 1

Evaluate the line integral: $I = \int_C F \cdot dr$

Where:

$$F(x,y) = \langle xy^2, -x^2 \rangle$$
$$r(t) = \langle t^3, t^2 \rangle, \quad 0 \le t \le 1$$

$\begin{array}{l} x = t^3 \\ y = t^2 \end{array}$	$\begin{array}{l} F(x,y) = \langle xy^2, -x^2 \rangle \\ F(x,y) = \langle (t^3)(t^2)^2, -(t^3)^2 \rangle \end{array}$

$$I = \int_C F \cdot dr = \int_0^1 F(r) \cdot (r') \ dt$$

$$I = \int_0^1 \langle (t^3)(t^2)^2, -(t^3)^2 \rangle \cdot \langle 3t^2, 2t \rangle \ dt$$

$$I = \int_0^1 \langle t^7, -t^6 \rangle \cdot \langle 3t^2, 2t \rangle \ dt$$

$$I = \int_0^1 (3t^9 - 2t^7) \ dt$$

$$I = \left[\frac{3}{10}t^{10} - \frac{2}{8}t^8 \right]_0^1 = \left[\frac{3}{10} - \frac{1}{4} \right]$$

$$I = \left[\frac{6}{20} - \frac{5}{20} \right] = \frac{1}{20}$$

(Stewart, Calculus Early Transcendentals, p. 1085, #19)

Line Integrals of Vector Fields -- Ex. 2a

Evaluate the line integral: $I = \int_C F \cdot dr$

Where: $F(x, y, z) = \langle \sin x, \cos y, xz \rangle$

$$r(t) = \langle t^3, -t^2, t \rangle, \quad 0 \le t \le 1$$

$\begin{aligned} x &= t^3 \\ y &= -t^2 \\ z &= t \end{aligned}$	$F(x, y, z) = \langle \sin x, \cos y, xz \rangle$ $F(x, y) = \langle \sin(t^3), \cos(-t^2), t^4 \rangle$

$I = \int_C F \cdot dr$

$I = \int_0^1 F(r) \cdot (r') \; dt$

$I = \int_0^1 \langle \sin t^3, \cos(-t^2), t^3 t \rangle \cdot \langle 3t^2, -2t, 1 \rangle \; dt$

$I = \int_0^1 \langle \sin t^3, \cos(-t^2), t^4 \rangle \cdot \langle 3t^2, -2t, 1 \rangle \; dt$

$I = \int_0^1 (3t^2 \sin t^3 - 2t \cos(-t^2) + t^4) \; dt$

(Stewart, Calculus Early Transcendentals, p. 1085, #21)

Line Integrals of Vector Fields -- Ex. 2b

Evaluate the line integral: $I = \int_C F \cdot dr$

Where: $F(x, y, z) = \langle \sin x, \cos y, xz \rangle$

$r(t) = \langle t^3, -t^2, t \rangle, \qquad 0 \le t \le 1$

$I = \int_0^1 (3t^2 \sin t^3 - 2t \cos(-t^2) + t^4) \, dt$

$\int_0^1 (3t^2 \sin t^3) \, dt$ U-Sub, $u = t^3$	$= [-\cos t^3]_0^1$ $= -[\cos 1 - \cos 0]$ $= -\cos 1 + 1$
$\int_0^1 -2t \cos(-t^2) \, dt$ U-Sub, $u = -t^2$	$= [\sin(-t^2)]_0^1$ $= \sin(-1) = -\sin(1)$
$\int_0^1 (t^4) \, dt$	$= \frac{1}{5}[t^5]_0^1 = \frac{1}{5}$

$I = [-\cos 1 + 1] - \sin 1 + \frac{1}{5}$

$I = -\cos 1 - \sin 1 + \frac{6}{5} \qquad \approx \qquad -0.182$

Desmos Check	$\int_0^1 \left(3t^2 \cdot \sin(t^3) - 2t \cdot \cos(-t^2) + t^4 \right) dt$ $ = -0.181773290676$

Line Integrals of Vector Fields -- Ex. 3

Evaluate the line integral of $F = \langle x, y, z \rangle$

Along the path: $c(t) = \langle t^2, 3t, 2t^2 \rangle$

With: $-1 \leq t \leq 2$

$F(c(t))$	$F(c(t)) = \langle t^2, 3t, 2t^2 \rangle$
$c'(t)$	$c'(t) = \langle 2t, 3, 4t \rangle$

$I = \int_C F \cdot dr$

$I = \int_{-1}^{2} F(c) \cdot (c') \, dt$

$I = \int_{-1}^{2} \langle t^2, 3t, 2t^2 \rangle \cdot \langle 2t, 3, 4t \rangle \, dt$

$I = \int_{-1}^{2} 2t^3 + 9t + 8t^3 \, dt$

$I = \int_{-1}^{2} 10t^3 + 9t \, dt$

$I = \left[\frac{10}{4} t^4 + \frac{9}{2} t^2 \right]_{-1}^{2}$

$I = \left[\left(\left(\frac{5}{2} \right) 16 + \left(\frac{9}{2} \right) 4 \right) - \left(\left(\frac{5}{2} \right) 1 + \left(\frac{9}{2} \right) 1 \right) \right]$

$I = \left[(40 + 18) - \left(\frac{5}{2} + \frac{9}{2} \right) \right] = 51$

Fundamental Theorem of Line Integrals

Fundamental Theorem of Line Integrals

The Fundamental Theorem of Line Integrals:

$$\int_C \nabla f \cdot dr \;=\; f(r(b)) - f(r(a))$$

Where ∇f is continuous on C

And $r = r(t), \quad a \le t \le b$

Recall the Fundamental Theorem of Calculus:

$$\int_a^b F'(x)\, dx \;=\; F(b) - F(a)$$

Where F' is continuous on $[\,a, b\,]$

Independence of Path
For a Conservative Vector Field

Given:	$F = Pi + Qj$
If:	$\dfrac{\delta P}{\delta y} = \dfrac{\delta Q}{\delta x}$
Then:	F is a conservative vector field.

$\int_C F \cdot dr$ is independent of path in D

IFF $\quad \int_C F \cdot dr = 0$

For every closed path C in D

Suppose:

C_1 and C_2 are two smooth curves (paths) that have

the same initial point A and terminal point B. Then:

$$\int_{C_1} \nabla f \cdot dr = \int_{C_2} \nabla f \cdot dr$$

The line integral of a conservative vector field

depends only on the initial point and terminal point of

a curve − **it is independent of path.**

Conservative Vector Field -- Ex. 1

Is the given vector field conservative?

Given: $F = \langle x - y, \ x - 3 \rangle$

$F = Pi + Qj$

$F = (x - y)i + (x - 3)j$

$P = x - y$	$Q = x - 3$
$\dfrac{\delta}{\delta y} P = \dfrac{\delta}{\delta y}(x - y)$	$\dfrac{\delta}{\delta x} Q = \dfrac{\delta}{\delta x}(x - 3)$
$\dfrac{\delta P}{\delta y} = \dfrac{\delta x}{\delta y} - \dfrac{\delta y}{\delta y}$	$\dfrac{\delta Q}{\delta x} = \dfrac{\delta x}{\delta x} - \dfrac{\delta}{\delta y}3$
$\dfrac{\delta P}{\delta y} = 0 - 1$	$\dfrac{\delta Q}{\delta x} = 1 - 0$
$\dfrac{\delta P}{\delta y} = -1$	$\dfrac{\delta Q}{\delta x} = 1$

$\dfrac{\delta P}{\delta y} \neq \dfrac{\delta Q}{\delta x}$

So, F is NOT a conservative vector field.

Conservative Vector Field -- Ex. 2

Is the given vector field conservative?

Given: $F = \langle\, 4 + 2xy,\ x^2 - 5y^2 \,\rangle$

$F = Pi + Qj$

$F = (4 + 2xy)i + (x^2 - 5y^2)j$

$P = 4 + 2xy$	$Q = x^2 - 5y^2$
$\dfrac{\delta P}{\delta y} = \dfrac{\delta}{\delta y}(4 + 2xy)$	$\dfrac{\delta Q}{\delta x} = \dfrac{\delta}{\delta x}(x^2 - 5y^2)$
$\dfrac{\delta P}{\delta y} = 0 + 2x$	$\dfrac{\delta Q}{\delta x} = 2x - 0$
$\dfrac{\delta P}{\delta y} = 2x$	$\dfrac{\delta Q}{\delta x} = 2x$

$$\frac{\delta P}{\delta y} = \frac{\delta Q}{\delta x}$$

So, F is a conservative vector field.

Conservative Vector Field -- Ex. 3

Find a function F such that $\nabla f = F$

Given: $F = \langle\, y^2,\ 2xy + e^{3z},\ 3ye^{3z}\,\rangle$

If there is such a function, then...	$f_x\,(x,y,z)\ =\ y^2$ $f_y\,(x,y,z)\ =\ 2xy + e^{3z}$ $f_z\,(x,y,z)\ =\ 3ye^{3z}$

$\int (f_x)\, dx\ =\ xy^2\ +\ f(y,z)$

$\int (f_y)\, dy\ =\ xy^2\ +\ ye^{3z}\ +\ f(x,z)$

$\int (f_z)\, dz\ =\ ye^{3z}\ +\ f(x,y)$

Compare above equations to get ...

$f(x,y,z)\ =\ xy^2\ +\ ye^{3z}\ +\ K$

$\nabla f\ =\ \langle\, f_x,\ f_y,\ f_z\,\rangle$

$\nabla f\ =\ \langle\, y^2,\ 2xy + e^{3z},\ 3ye^{3z}\,\rangle\ =\ F$

(Stewart, Calculus Early Transcendentals, p. 1092)

Conservative Vector Field -- Ex. 4a

Evaluate: $\int_{c_1} F \, ds$ and $\int_{c_2} F \, ds$

From point $(1,1,1)$ to point $(1,2,4)$

Given: $F = \langle 2xyz, \, x^2z, \, x^2y \rangle$

$c_1 = \langle 1, t, t^2 \rangle$ and $c_2 = \langle 1, t+1, 3t+1 \rangle$

If the integrals are equal, explain your answer

$\int_{c_1} F \, ds \; = \; \int_{t_1}^{t_2} F(c_1) \cdot c_1' \; dt$

$\qquad F(c_1) = \langle 2t^3, t^2, t \rangle$

$\qquad c_1 = \langle 1, t, t^2 \rangle$ and $c_1' = \langle 0, 1, 2t \rangle$

$\int_{c_1} F \, ds \; = \; \int_{t_1}^{t_2} \langle 2t^3, t^2, t \rangle \cdot \langle 0, 1, 2t \rangle \; dt$

$\int_{c_1} F \, ds \; = \; \int_{t_1}^{t_2} (t^2 + 2t^2) \; dt$

$\qquad$ point $(1,1,1)$ to $(1,2,4)$

$\qquad \rightarrow t_1 = 1$ and $t_2 = 2$

$\int_{c_1} F \, ds \; = \; \int_1^2 (3t^2) \; dt \; = \; [t^3]_1^2 \; = \; 7$

Conservative Vector Field -- Ex. 4b

Evaluate: $\int_{c_1} F \, ds$ and $\int_{c_2} F \, ds$

From point $(1,1,1)$ to point $(1,2,4)$

Given: $F = \langle 2xyz, \; x^2 z, \; x^2 y \rangle$

$c_1 = \langle 1, t, t^2 \rangle$ and $c_2 = \langle 1, t+1, 3t+1 \rangle$

If the integrals are equal, explain your answer

$\int_{c_2} F \, ds = \int_{t_1}^{t_2} F(c_2) \cdot c_2' \, dt$

$F(c_2) = \langle 6t^2 + 8t + 2, \; 3t + 1, \; t + 1 \rangle$

$c_2' = \langle 0, \; 1, \; 3 \rangle$

$\int_{c_2} F \, ds = \int_{t_1}^{t_2} F(c_2) \cdot \langle 0, 1, 3 \rangle \, dt$

$\int_{c_2} F \, ds = \int_{t_1}^{t_2} (6t + 4) \, dt$

point $(1,1,1)$ to $(1,2,4)$

$\rightarrow t_1 = 0$ and $t_2 = 1$

$\int_{c_2} F \, ds = \int_{0}^{1} (6t + 4) \, dt = [\, 3t^2 + 4 \,]_0^1 = 7$

Conservative Vector Field -- Ex. 4c

Evaluate: $\int_{c_1} F \, ds$ and $\int_{c_2} F \, ds$

From point $(1, 1, 1)$ to point $(1, 2, 4)$

Given: $F = \langle 2xyz, x^2z, x^2y \rangle$

$c_1 = \langle 1, t, t^2 \rangle$ and $c_2 = \langle 1, t + 1, 3t + 1 \rangle$

If the integrals are equal, explain your answer

Previously Found:

$\int_{c_1} F \, ds = \int_1^2 (3t^2) \, dt = [t^3]_1^2 = 7$

$\int_{c_2} F \, ds = \int_0^1 (6t + 4) \, dt = [3t^2 + 4]_0^1 = 7$

Both integrals are equal because line integrals are

independent of path if both paths are in the domain.

Green's Theorem

Green's Theorem

Green's Theorem gives the relationship between a line integral around a closed curve C and a double integral over the region D bounded by C.

Positive Orientation

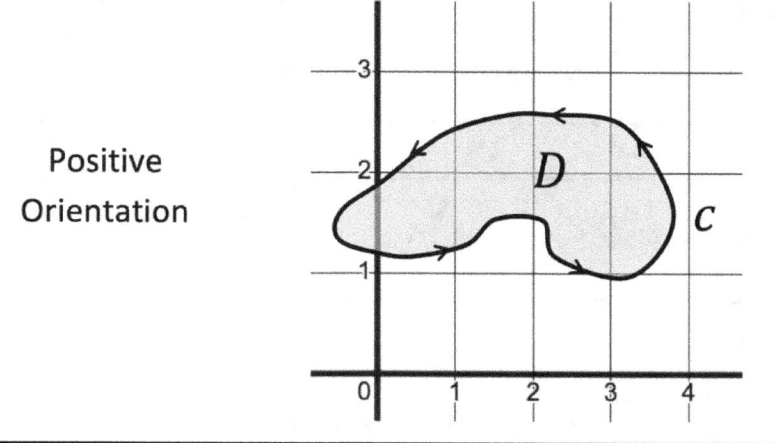

Green's Theorem: Let C be a positively oriented, piecewise smooth, simple closed curve in a plane, and let D be the retion bounded by C. If P and Q are functions of (x, y) defined on an open region containing D and have continuous partial derivatives, then:

$$\oint_C P\, dx + Q\, dy = \iint_D \left(\frac{\delta Q}{\delta x} - \frac{\delta P}{\delta y} \right) dA$$

Green's Theorem -- Ex. 1

Evaluate

$\int_C x^5 \, dx + x^2 y \, dy$

Where C is the triangular curve going counter-clockwise from: $(0,0)$ to $(1,0)$ to $(0,1)$ to $(0,0)$

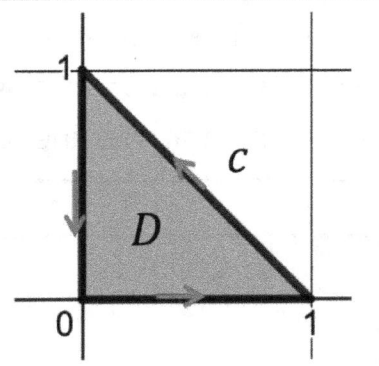

$\oint_C P \, dx + Q \, dy = \iint_D \left(\frac{\delta Q}{\delta x} - \frac{\delta P}{\delta y} \right) dA$

$\oint_C x^5 \, dx + x^2 y \, dy$

$= \int_0^1 \int_0^{1-x} (2xy - 0) \; dy \, dx$

$= \int_0^1 2x \left[\frac{1}{2} y^2 \right]_0^{1-x} \; dx$

$= \int_0^1 x \left[(1 - x)^2 \right] \; dx$

$= \int_0^1 x \left[1 - 2x + x^2 \right] \; dx$

$= \int_0^1 (x - 2x^2 + x^3) \; dx$

$= \left[\frac{1}{2} x^2 - \frac{2}{3} x^3 + \frac{1}{4} x^4 \right]_0^1 = \frac{1}{12} \approx 0.083$

Green's Theorem -- Ex. 2

Evaluate

$\int_C (3y - e^x)\, dx\ +\ (7x + y)\, dy$

Where C is the circle: $x^2 + y^2 = 9$

$\oint_C P\, dx\ +\ Q\, dy\ =\ \iint_D \left(\frac{\delta Q}{\delta x} - \frac{\delta P}{\delta y} \right) dA$

$\int_C (3y - e^x)\, dx\ +\ (7x + y)\, dy$

$=\ \iint_D \left[\frac{\delta}{\delta x}(7x + y)\ -\ \frac{\delta}{\delta y}(3y - e^x) \right] dA$

$=\ \iint_D [\, 7\ -\ 3\,]\, dA\ \ =\ \ \iint_D 4\ dA$

$=\ 4 \int_0^{2\pi} \int_0^3\ r\ dr\ d\theta$

$=\ 4 \int_0^{2\pi} \left[\frac{1}{2} r^2 \right]_0^3\ d\theta$

$=\ 2 \int_0^{2\pi} [9]\ d\theta$

$=\ 18\, [\, \theta\,]_0^{2\pi}\ \ =\ 18\, [\, 2\pi\,]\ =\ 36\pi\ \ \approx\ \ 113.1$

Green's Theorem -- Ex. 3a

Evaluate the line integral by two methods:
(a.) Directly, and (b.) Using Green's Theorem.
Given: $F = \langle x^2 - y, y^3 \rangle$ and C is the unit
circle: $x^2 + y^2 = 1$, counter-clockwise.

Part (a.) Direct evaluation of line integral

Polar Coord. $r = 1$	$x = \cos t \;\rightarrow\; dx = -\sin t\, dt$ $y = \sin t \;\rightarrow\; dy = \cos t\, dt$

$\int_C = \int (x^2 - y)dx + (y^2)\, dy$

$\int_C = \int_0^{2\pi} (\cos^2 t - \sin t)(-\sin t)$
$\qquad + \;\; (\sin^3 t)(\cos t)\; dt$

$\int_C = \int_0^{2\pi} (-\cos^2 t \cdot \sin t + \sin^2 t)$
$\qquad + \;\; (\sin^3 t \cdot \cos t)\; dt$

$\int_C = \int_0^{2\pi} \left(\cos^2 t \right)(-\sin t) + \frac{1}{2}(1 - \cos 2t) \right)$
$\qquad + \;\; (\sin^3 t \cdot \cos t)\; dt$

$\int_C = \left[\frac{1}{3}\cos^3 t \; + \; \frac{1}{2}t \; - \; \frac{\sin 2t}{2} \; + \; \frac{1}{4}\sin^4 t \right]_0^{2\pi}$

$\int_C = \left[\left(\frac{1}{3} + \pi - 0 + 0 \right) - \left(\frac{1}{3} + 0 + 0 + 0 \right) \right] \; = \; \pi$

Green's Theorem -- Ex. 3b

Evaluate the line integral by two methods:
(a.) Directly, and (b.) Using Green's Theorem.
Given: $F = \langle x^2 - y, y^3 \rangle$ and C is the unit
circle: $x^2 + y^2 = 1$, counter-clockwise.

Part (a.) Direct evaluation of line integral	
Previously Found	$\int_C = \pi$

Part (b.) Use Green's Theorem	
$P = x^2 - y$	$Q = y^3$
$\dfrac{\delta P}{\delta y} = -1$	$\dfrac{\delta Q}{\delta x} = 0$

$$\oint_C P \, dx + Q \, dy = \iint_D \left(\frac{\delta Q}{\delta x} - \frac{\delta P}{\delta y} \right) dA$$

$$\int_C = \int_0^{2\pi} \int_0^1 (0 - (-1)) \; r \, dr \, d\theta$$

$$\int_C = \int_0^{2\pi} \int_0^1 (1) \; r \, dr \, d\theta$$

Same
Answer!

$$\int_C = \int_0^{2\pi} \int_0^1 (r) \; dr \, d\theta$$

$$\int_C = \int_0^{2\pi} \left[\frac{r^2}{2} \right]_0^1 d\theta = 2\pi \left(\tfrac{1}{2} \right) = \pi$$

Green's Theorem -- Ex. 4a

Evaluate the line integral by two methods:

(a.) Directly, and

(b.) Using Green's Theorem.

Given: $F = \langle xy, x^2 \rangle$ and

C is a rectangle with vertices:

$(0,0), (2,0), (2,1), (0,1)$

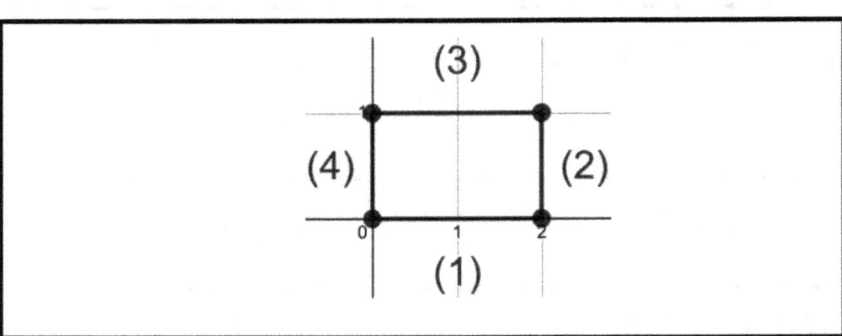

Green's Theorem -- Ex. 4b

	Part (a.) Direct evaluation of line integral
C1	$(0,0) \rightarrow (2,0) \qquad \vec{u} = \langle 2,0 \rangle$ $L(t) = (0,0) + t(2,0) = \langle 2t, 0 \rangle$ $x(t) = 2t, \quad dx = 2\, dt$ $y(t) = 0, \qquad dy = 0\, dt$
	$\int_{C_1} xy\, dx + x^2\, dy$ $= \int_0^1 (2t)(0)(2) + (2t)(0)\, dt = 0$
C2	$(0,2) \rightarrow (2,1) \qquad \vec{u} = \langle 0,1 \rangle$ $L(t) = (2,0) + t(0,1) = \langle 2, t \rangle$ $x(t) = 2, \quad dx = 0\, dt$ $y(t) = t, \qquad dy = 1\, dt$
	$\int_{C_2} xy\, dx + x^2\, dy = \int_{C_2} x^2\, dy$ $= \int_0^1 (2)^2 (1)\, dt = 4$

Green's Theorem -- Ex. 4c

	Part (a.) Direct evaluation of line integral (Continued)
C3	$(2,1) \rightarrow (0,1) \qquad \vec{u} = \langle -2,0 \rangle$ $L(t) = (2,1) + t(-2,0) = \langle 2 - 2t, 1 \rangle$ $x(t) = 2 - 2t, \quad dx = -2dt$ $y(t) = 1, \qquad\quad dy = 0 \; dt$
	$\int_{C_3} xy \, dx + x^2 \, dy = \int_{C_2} xy \, dx$ $\quad = \int_0^1 (2 - 2t)(1)(-2) \, dt = -2$
C4	$(0,1) \rightarrow (0,0) \qquad \vec{u} = \langle 0, -1 \rangle$ $L(t) = (0,1) + t(0, -1) = \langle 0, 1 - t \rangle$ $x(t) = 0, \qquad\quad dx = 0 \; dt$ $y(t) = 1 - t, \quad dy = -1 \; dt$
	$\int_{C_4} xy \, dx + x^2 \, dy = 0$

Green's Theorem -- Ex. 4d

Part (a.) Direct evaluation of line integral

Previously Found	$\int_C = \int_{C_1} + \int_{C_2} + \int_{C_3} + \int_{C_4}$
	$\int_C = 0 + 4 - 2 + 0 = 2$

Part (b.) Green's Theorem

$P = xy$	$Q = x^2$
$\dfrac{\delta P}{\delta y} = x$	$\dfrac{\delta Q}{\delta x} = 2x$

$$\oint_C P\,dx + Q\,dy = \iint_D \left(\frac{\delta Q}{\delta x} - \frac{\delta P}{\delta y} \right) dA$$

$$\int_C = \int_0^1 \int_0^2 (2x - x)\,dx\,dy$$

$$\int_C = \int_0^1 \int_0^2 (x)\,dx\,dy \qquad \text{Same Answer!}$$

$$\int_C = \int_0^1 \left[\frac{1}{2} x^2 \right]_0^2 dy = \int_0^1 [2]\,dy = 2$$

Green's Theorem -- Ex. 5

Evaluate: $\int_C x^2\, dx - xy\, dy + dz$

Where: C is the parabola $z = y^2,\ y = 0$

From: $(-1,0,1)$ to $(1,0,1)$

Parameterize the path on the parabola. $z = y^2$	Let: $\quad x = t$ Then: $z = t^2$ and $y = 0$ And: $\quad C(t) = \langle t, t^2, 0 \rangle$
Find t_1 and t_2	From $(-1,0,1)$ to $(1,0,1)$ $\rightarrow \quad t_1 = -1$ and $t_2 = 1$

$\int_C x^2\, dx - xy\, dy + dz$

$= \int_C \langle x^2, -xy, 1 \rangle \cdot \langle dx, dy, dz \rangle$

$=. \int_{-1}^{1} \langle t^2, -(t)(0), 1 \rangle \cdot \langle 1, 0, 2t \rangle\ dt$

$= \int_{-1}^{1} \langle t^2, 0, 1 \rangle \cdot \langle 1, 0, 2t \rangle\ dt$

$= \int_{-1}^{1} (t^2 + 2t)\, dt \ = \ \left[\dfrac{t^3}{3} + t^2 \right]_{-1}^{1}$

$= \left(\dfrac{1}{3} + 1 \right) - \left(-\dfrac{1}{3} + 1 \right) \ = \ \dfrac{2}{3}$

Green's Theorem -- Ex. 6a

Evaluate: $\int_C (y - x)\, dx + (xy)\, dy$

Where C is the boundary of the region D
in the upper-half plane between the circles

$$x^2 + y^2 = 1 \quad \text{and} \quad x^2 + y^2 = 4$$

Polar Coord:

$x = r \cos \theta$

$y = r \sin \theta$

$1 \le r \le 2$

$0 \le \theta \le \pi$

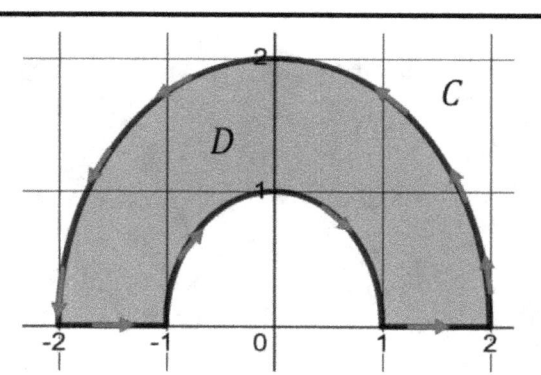

Conservative Check

$$\frac{\delta P}{\delta y} = \frac{\delta}{\delta y}(y - x) = 1$$

$$\frac{\delta Q}{\delta x} = \frac{\delta}{\delta x}(xy) = y$$

$$\frac{\delta P}{\delta y} \ne \frac{\delta Q}{\delta x} \rightarrow \text{Not Conservative}$$

Green's Theorem:

$$\int_C P\, dx + Q\, dy = \iint_D \left(\frac{\delta Q}{\delta x} - \frac{\delta P}{\delta y} \right) dA$$

$$\int_C (y - x)\, dx + (xy)\, dy = \iint_D (y - 1)\, dA$$

Green's Theorem -- Ex. 6b

Evaluate: $\int_C (y - x)\,dx + (xy)\,dy$

Where C is the boundary of the region D
in the upper-half plane between the circles

$$x^2 + y^2 = 1 \quad \text{and} \quad x^2 + y^2 = 4$$

Polar Coord:

$x = r \cos \theta$

$y = r \sin \theta$

$1 \leq r \leq 2$

$0 \leq \theta \leq \pi$

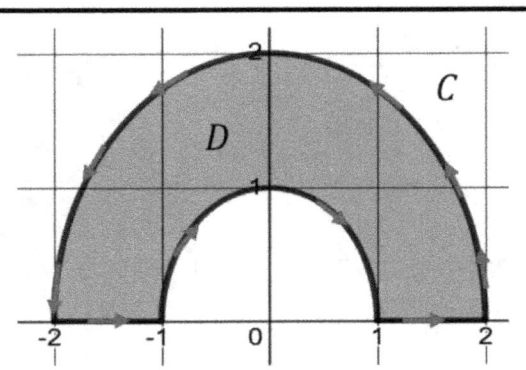

Green's Theorem:

$$\int_C P\,dx + Q\,dy \;=\; \iint_D \left(\frac{\delta Q}{\delta x} - \frac{\delta P}{\delta y} \right)\, dA$$

$$\int_C (y - x)\,dx + (xy)\,dy \;=\; \iint_D (y - 1)\, dA$$

$$= \int_0^\pi \int_1^2 (r \sin \theta - 1)\; r\,dr\,d\theta$$

$$= \int_0^\pi \int_1^2 (r^2 \sin \theta - r)\; dr\,d\theta$$

$$= \int_0^\pi \left[\frac{1}{3} r^3 \sin \theta \;-\; \frac{1}{2} r^2 \right]_1^2 \; d\theta$$

Green's Theorem -- Ex. 6c

Evaluate: $\int_C (y-x)\,dx + (xy)\,dy$

Where C is the boundary of the region D
in the upper-half plane between the circles

$$x^2 + y^2 = 1 \quad \text{and} \quad x^2 + y^2 = 4$$

Polar Coord:

$x = r\cos\theta$
$y = r\sin\theta$
$1 \le r \le 2$
$0 \le \theta \le \pi$

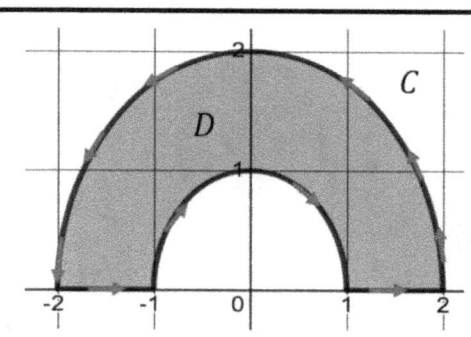

$\int_C (y-x)\,dx + (xy)\,dy$

$$= \int_0^\pi \left[\frac{1}{3}r^3 \sin\theta - \frac{1}{2}r^2 \right]_1^2 d\theta$$

$$= \int_0^\pi \left[\left(\frac{8}{3}\sin\theta - 2 \right) - \left(\frac{1}{3}\sin\theta - \frac{1}{2} \right) \right] d\theta$$

$$= \int_0^\pi \left(\frac{7}{3}\sin\theta - \frac{3}{2} \right) d\theta$$

$$= \left[-\frac{7}{3}\cos\theta - \frac{3}{2}\theta \right]_0^\pi$$

$$= \left[\left(\frac{7}{3} - \frac{3}{2}\pi \right) - \left(-\frac{7}{3} \right) \right] = \frac{14}{3} - \frac{3}{2}\pi$$

Curl and Divergence

Curl -- Given: $F = \langle P, Q, R \rangle$

$$curl \; F \; = \; \left\langle \left(\frac{\delta R}{\delta y} - \frac{\delta Q}{\delta z}\right), \left(\frac{\delta P}{\delta z} - \frac{\delta R}{\delta x}\right), \left(\frac{\delta Q}{\delta x} - \frac{\delta P}{\delta y}\right) \right\rangle$$

$$curl \; F \; = \; \langle (R_y - Q_z), (P_z - R_x), (Q_x - P_y) \rangle$$

$$curl \; F \; = \; \nabla \times F$$

$$curl \; F \; = \; \begin{vmatrix} i & j & k \\ \dfrac{\delta}{\delta x} & \dfrac{\delta}{\delta y} & \dfrac{\delta}{\delta z} \\ P & Q & R \end{vmatrix}$$

The curl of a gradient vector field is 0.

$$curl \; (\nabla F) = 0$$

Recall: $\nabla F = \left\langle \dfrac{\delta P}{\delta x}, \dfrac{\delta Q}{\delta y}, \dfrac{\delta R}{\delta z} \right\rangle = \langle P_x, Q_y, R_z \rangle$

If F is conservative, then: $curl \; F = 0$

Recall: Conservative vector field has: $F = \nabla f$

Divergence -- Given: $F = \langle P, Q, R \rangle$
$div\ F\ =\ \dfrac{\partial P}{\partial x} + \dfrac{\partial Q}{\partial y} + \dfrac{\partial R}{\partial z}$
$div\ F\ =\ \nabla \cdot F$ $\qquad\qquad$ (Scalar)

Recall:

$$curl\ F\ =\ \nabla \times F \qquad\qquad \text{(Vector)}$$

$$curl\ F\ =\ \begin{vmatrix} i & j & k \\ \dfrac{\delta}{\delta x} & \dfrac{\delta}{\delta y} & \dfrac{\delta}{\delta z} \\ P & Q & R \end{vmatrix}$$

$$div\ curl\ F\ =\ 0$$

Curl -- Ex. 1

Given: $F = \langle xz,\ xyz,\ y \rangle$ Find: $curl\ F$

$F = \langle P, Q, R \rangle = \langle xz,\ xyz,\ y \rangle$

$curl\ F\ =\ \nabla \times F$

$$curl\ F\ =\ \begin{vmatrix} i & j & k \\ \frac{\delta}{\delta x} & \frac{\delta}{\delta y} & \frac{\delta}{\delta z} \\ P & Q & R \end{vmatrix} = \begin{vmatrix} i & j & k \\ \frac{\delta}{\delta x} & \frac{\delta}{\delta y} & \frac{\delta}{\delta z} \\ xz & xyz & y \end{vmatrix}$$

$$curl\ F\ =\ +\left[\frac{\delta}{\delta y}(y) - \frac{\delta}{\delta z}(xyz) \right] i$$

$$-\left[\frac{\delta}{\delta x}(y) - \frac{\delta}{\delta z}(xz) \right] j$$

$$+\left[\frac{\delta}{\delta x}(xyz) - \frac{\delta}{\delta y}(xz) \right] k$$

$curl\ F\ =\ [\,1 - xy\,]\,i - [\,0 - x\,]\,j + [\,yz - 0\,]\,k$

$curl\ F\ =\ [\,1 - xy\,]\,i + [\,x\,]\,j + [\,yz\,]\,k$

$curl\ F\ =\ \langle\, 1 - xy, x, yz \,\rangle$

Curl -- Ex. 2

Show that F is not a conservative vector.

Given: $F = \langle xz, \ xyz, \ y \rangle$

In the previous example (#1) we showed:

$$curl \ F = \langle 1 - xy, x, yz \rangle$$

F is NOT conservative because: $curl \ F \neq 0$

If F is conservative, then: $curl \ F = 0$

Curl -- Ex. 3a

Show that F is a conservative vector field and find a function, such that, $F = \nabla f$

Given: $F = \langle y^2z^3, \ 2xyz^3, \ 3xy^2z^2 \rangle$

$\text{curl } F = \nabla \times F$

$$= \begin{vmatrix} i & j & k \\ \dfrac{\delta}{\delta x} & \dfrac{\delta}{\delta y} & \dfrac{\delta}{\delta z} \\ y^2z^3 & 2xyz^3 & 3xy^2z^2 \end{vmatrix}$$

$$= + \left[\frac{\delta}{\delta y}(3xy^2z^2) - \frac{\delta}{\delta z}(2xyz^3) \right] i$$

$$- \left[\frac{\delta}{\delta x}(3xy^2z^2) - \frac{\delta}{\delta z}(y^2z^3) \right] j$$

$$+ \left[\frac{\delta}{\delta x}(2xyz^3) - \frac{\delta}{\delta y}(y^2z^3) \right] k$$

$$= + [\, 6xyz^2 - 6xyz^2 \,] i$$

$$- [\, 3y^2z^2 - 3y^2z^2 \,] j$$

$$+ [\, 2yz^3 - 2yz^3 \,] k$$

$\text{curl } F = 0 \qquad$ Therefore, F is conservative.

(Stewart, Calculus Early Transcendentals, p. 1105)

Curl -- Ex. 3b

Show that F is a conservative vector field and find a

function, such that, $F = \nabla f$

Given: $F = \langle y^2z^3, \; 2xyz^3, \; 3xy^2z^2 \rangle$

Previously, we found F is conservative because:

$curl \; F = 0$

Now, find a function f , such that, $F = \nabla f$

Here: $F = \nabla f = \langle f_x, f_y, f_z \rangle$

$f_x = y^2z^3$ $\qquad$ $f_y = 2xyz^3$ $\qquad$ $f_z = 3xy^2z^2$

$\int y^2z^3 \; dx \qquad \rightarrow \quad f = xy^2z^3 + f(y,z)$

$\int 2xyz^3 \; dy \qquad \rightarrow \quad f = xy^2z^3 + f(x,z)$

$\int 3xy^2z^2 \; dz \quad \rightarrow \quad f = xy^2z^3 + f(x,y)$

Using the above three equations ...

$$f = xy^2z^3 + K$$

Divergence -- Ex. 4

Given: $F = \langle xz, \; xyz, \; y \rangle$

Find: $div\, F$ and $div\, curl\, F$

$div\, F = \nabla \cdot F = \dfrac{\partial P}{\partial x} + \dfrac{\partial Q}{\partial y} + \dfrac{\partial R}{\partial z}$

$div\, F = z + xz + 0 = z + xz$

Previously found: $curl\, F = \langle 1 - xy, x, yz \rangle$

$div\, curl\, F = div\, \langle 1 - xy, x, yz \rangle$

$= \dfrac{\partial}{\partial x}(1 - xy) + \dfrac{\partial}{\partial y}(x) + \dfrac{\partial}{\partial z}(yz)$

$= (0 - y) + (0) + (y)$

$= -y + y$

$= 0$

Parametric Surfaces and Areas

Parametric Surfaces & Parametric Areas

$A(S) =$ Surface Area of S

$$A(S) = \iint_D |r_u \times r_v| \, dA \quad ; \quad (u, v) \in D$$

Where:

$r(u, v) =$ Vector equation of a surface.

$r(u, v) = \langle\, x(u, v),\ y(u, v\,),\ z(u, v)\,\rangle$

$$r_u = \langle \frac{\delta x}{\delta u}, \frac{\delta y}{\delta u}, \frac{\delta z}{\delta u} \rangle \quad ; \quad r_v = \langle \frac{\delta x}{\delta v}, \frac{\delta y}{\delta v}, \frac{\delta z}{\delta v} \rangle$$

Surface Area of Graph of Function
Special Case: $z = f(x, y)$

$A(S)$ = Surface Area of S

$$A(S) = \iint_D \sqrt{1 + \left(\frac{\delta z}{\delta x}\right)^2 + \left(\frac{\delta z}{\delta y}\right)^2} \, dA$$

Where:

$x = x$ $\qquad$ $y = y$ $\qquad$ $z = f(x, y)$

$(x, y) \in D$

Parametric Surface Area -- Ex. 1a

Find the surface area of a sphere of radius a.

Given: $x^2 + y^2 + z^2 = a^2$

Parametric Representation: $\rho = r = a$

$x = a \cos\theta \sin\phi$

$y = a \sin\theta \sin\phi$

$z = a \cos\phi$

$r(\phi,\theta) = \langle a \cos\theta \sin\phi, a \sin\theta \sin\phi, a \cos\phi \rangle$

$D = $ Domain

$D = \{(\phi,\theta) \mid 0 \le \phi \le \pi, \ 0 \le \theta \le 2\pi\}$

Cross Product of the tangent vectors:

$$r_\phi \times r_\theta = \begin{vmatrix} i & j & k \\ \dfrac{\delta x}{\delta \phi} & \dfrac{\delta y}{\delta \phi} & \dfrac{\delta z}{\delta \phi} \\ \dfrac{\delta x}{\delta \theta} & \dfrac{\delta y}{\delta \theta} & \dfrac{\delta z}{\delta \theta} \end{vmatrix}$$

(Stewart, Calculus Early Transcendentals, p. 1117)

Parametric Surface Area -- Ex. 1b

Find the surface area of a sphere of radius a.

Given: $x^2 + y^2 + z^2 = a^2$

$r_\phi \times r_\theta =$

$$\begin{vmatrix} i & j & k \\ a \cos\theta \, \cos\phi & a \sin\theta \, \cos\phi & -a \sin\phi \\ -a \sin\theta \, \sin\phi & a \cos\theta \, \sin\phi & 0 \end{vmatrix}$$

$r_\phi \times r_\theta = (a^2 \cos\theta \, \sin^2\phi) \, i$

$+ (a^2 \sin\theta \, \sin^2\phi) \, j$

$+ (a^2 \cos\theta \, \sin\phi) \, k$

$|r_\phi \times r_\theta| = \sqrt{()^2 + ()^2 + ()^2}$

$|r_\phi \times r_\theta| = \sqrt{\begin{array}{l} a^4 \cos^2\theta \, \sin^4\phi \; + \\ a^4 \sin^2\theta \, \sin^4\phi \; + \\ a^4 \cos^2\theta \, \sin^2\phi \end{array}}$

$|r_\phi \times r_\theta| = a^2\sqrt{\sin^2\phi} = a^2 \sin\phi$

Parametric Surface Area -- Ex. 1c

Find the surface area of a sphere of radius a.

Given: $x^2 + y^2 + z^2 = a^2$

Previously found:

$$|r_\phi \times r_\theta| = a^2\sqrt{\sin^2\phi} = a^2\sin\phi$$

$$A(S) = \iint_D |r_\phi \times r_\theta|\ dA$$

$$A(S) = \int_0^{2\pi} \int_0^{\pi} (a^2\sin\phi)\ d\phi\ d\theta$$

$$A(S) = a^2 \int_0^{2\pi} [-\cos\phi]_0^{\pi}\ d\theta$$

$$A(S) = a^2 \int_0^{2\pi} -[\cos\pi - \cos 0]\ d\theta$$

$$A(S) = a^2 \int_0^{2\pi} -[(-1) - (1)]\ d\theta$$

$$A(S) = a^2 \int_0^{2\pi} 2\ d\theta = 2a^2 [\theta]_0^{2\pi}$$

$$A(S) = 2a^2 [2\pi - 0] = 4\pi a^2$$

Parametric Surface Area -- Ex. 2a

Find the surface area of the part of the paraboloid

$z = x^2 + y^2$ that lies

under the plane $z = 9$

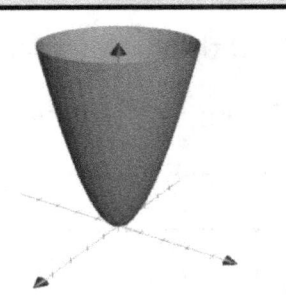

Note: This is a **special case where** $z = f(x, y)$

So use the equation:

$$A(S) = \iint_D \sqrt{1 + \left(\frac{\delta z}{\delta x}\right)^2 + \left(\frac{\delta z}{\delta y}\right)^2} \ dA$$

$$A(S) = \iint_D \sqrt{1 + (2x)^2 + (2y)^2} \ dA$$

$$A(S) = \iint_D \sqrt{1 + 4(x^2 + y^2)} \ dA$$

Convert to Polar Coordinates:

$$A = \int_0^{2\pi} \int_0^3 \sqrt{1 + 4r^2} \ r \ dr \ d\theta$$

$$A = \int_0^{2\pi} d\theta \int_0^3 \sqrt{1 + 4r^2} \ r \ dr$$

(Stewart, Calculus Early Transcendentals, p. 1118)

Parametric Surface Area -- Ex. 2b

Find the surface area of the
part of the paraboloid
$z = x^2 + y^2$ that lies
under the plane $z = 9$

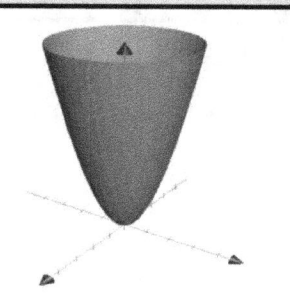

U-Sub: $u = 1 + 4r^2$ $du = 8r\ dr$	$\int_0^3 \sqrt{1 + 4r^2}\ r\ dr$ $= \frac{1}{8} \int u^{\frac{1}{2}}\ du\ =\ \frac{1}{8} \cdot \frac{2}{3} \left[u^{\left(\frac{3}{2}\right)} \right]$ $= \frac{1}{12} \left[u^{\left(\frac{3}{2}\right)} \right]$

$\int_0^3 \sqrt{1 + 4r^2}\ r\ dr\ =\ \frac{1}{12} \left[(1 + 4r^2)^{\frac{3}{2}} \right]_0^3$

$=\ \frac{1}{12} \left[(1 + 36)^{\frac{3}{2}} - (1)^{\frac{3}{2}} \right]\ =\ \frac{1}{12} \left[37\sqrt{37} - 1 \right]$

$A\ =\ \int_0^{2\pi} d\theta\ \int_0^3 \sqrt{1 + 4r^2}\ r\ dr$

$A\ =\ (2\pi) \left(\frac{1}{12} \right) \left[37\sqrt{37} - 1 \right]$

$A\ =\ \frac{\pi}{12} \left[37\sqrt{37} - 1 \right]$

Surface Integrals

Surface Integrals
For Parametric Surfaces

$\iint_S f(x,y,z)$ = Surface Integral of f over S

$\iint_S f(x,y,z)$ = $\iint_D f\left(r(u,v)\right) \, |\, r_u \times r_v\,|\, dA$

Where:

The surface has the vector equation
$r(u,v) = \langle\, x(u,v),\ y(u,v),\ z(u,v)\, \rangle$

Surface Integrals
For Functions With: $z = g(x,y)$

$\iint_S f(x,y,z) \, dS$

$$= \iint_D f\left(x,y,g(x,y)\right) \sqrt{\left(\frac{\delta z}{\delta x}\right)^2 + \left(\frac{\delta z}{\delta y}\right)^2 + 1} \ \ dA$$

Note: Either $x,\ y,\ z$
can be functions of the other two.

Surface Integrals
For Parametric Surfaces -- Ex. 1a

Compute the surface integral: $\iint_S x^2 \, dS$

Where S is a unit sphere: $\quad x^2 + y^2 + z^2 = 1$

Use parametric representation: With $r = 1$

$$x = \cos\theta \sin\phi \qquad y = \sin\theta \sin\phi \qquad z = \cos\phi$$

$$0 \le \theta \le 2\pi \qquad\qquad 0 \le \phi \le \pi$$

$$r(\theta, \phi) = \langle \cos\theta \sin\phi, \sin\theta \sin\phi, \cos\phi \rangle$$

$$\left| r_\phi \times r_\theta \right| = \begin{vmatrix} i & j & k \\ \dfrac{\delta x}{\delta \phi} & \dfrac{\delta y}{\delta \phi} & \dfrac{\delta z}{\delta \phi} \\ \dfrac{\delta x}{\delta \theta} & \dfrac{\delta y}{\delta \theta} & \dfrac{\delta z}{\delta \theta} \end{vmatrix}$$

$$= \begin{vmatrix} i & j & k \\ \cos\theta \cos\phi & \sin\theta \cos\phi & -\sin\phi \\ -\sin\theta \sin\phi & \cos\theta \sin\phi & 0 \end{vmatrix}$$

$$= \langle \cos\theta \, \sin^2\phi, \, \sin\theta \, \sin^2\phi, \, \sin\theta \, \sin\phi \rangle$$

(Stewart, Calculus Early Transcendentals, p. 1123)

Surface Integrals

For Parametric Surfaces -- Ex. 1b

Compute the surface integral: $\iint_S x^2 \, dS$

Where S is a unit sphere: $x^2 + y^2 + z^2 = 1$

$|r_\phi \times r_\theta| =$

$= \langle \cos\theta \, \sin^2\phi, \, \sin\theta \, \sin^2\phi, \, \sin\phi \, \cos\phi \rangle$

$= \sqrt{()^2 + ()^2 + ()^2} =$

$\sqrt{\cos^2\theta \sin^4\phi + \sin^2\theta \sin^4\phi + \sin^2\theta \cos^2\phi}$

$\sqrt{\sin^4\phi + \sin^2\phi \cos^2\phi} \quad = \quad \sqrt{\sin^2\phi} \quad = \quad \sin\phi$

$\iint_S f(x,y,z) \quad = \quad \iint_D f\left(r(\phi,\theta)\right) |r_\phi \times r_\theta| \, dA$

$= \int_0^{2\pi} \int_0^{\pi} (\cos\theta \sin\phi)^2 \, (\sin\phi) \, d\phi \, d\theta$

$= \int_0^{2\pi} \cos^2\theta \, d\theta \int_0^{\pi} \sin^3\phi \, d\phi$

Surface Integrals
For Parametric Surfaces -- Ex. 1c

Compute the surface integral: $\iint_S x^2 \ dS$

Where S is a unit sphere: $\quad x^2 + y^2 + z^2 = 1$

$\iint_S f(x,y,z) \ = \ \int_0^{2\pi} \cos^2 \theta \ d\theta \ \int_0^\pi \sin^3 \phi \ d\phi$	
$\int_0^{2\pi} \cos^2 \theta \ d\theta$	$\int_0^{2\pi} \frac{1}{2}(1 + \cos 2\theta) \ d\theta$ $\frac{1}{2} \int_0^{2\pi} (1 + \cos 2\theta) \ d\theta$ $\frac{1}{2} \left[\theta + \frac{1}{2} \sin 2\theta \right]_0^{2\pi}$ $\frac{1}{2} \left[\theta + \frac{1}{2} \sin 2\theta \right]_0^{2\pi}$ $\frac{1}{2} [2\pi] \ = \ \pi$
$\int_0^\pi \sin^3 \phi \ d\phi$	$\int_0^\pi (1 - \cos^2 \phi) \sin \phi \ d\phi$ $\int_0^\pi (\sin \phi - \cos^2 \phi \sin \phi) \, d\phi$ Continued ...

Surface Integrals
For Parametric Surfaces -- Ex. 1d

Compute the surface integral: $\iint_S x^2 \, dS$

Where S is a unit sphere: $\qquad x^2 + y^2 + z^2 = 1$

$$\iint_S f(x,y,z) \;=\; \int_0^{2\pi} \cos^2 \theta \, d\theta \int_0^{\pi} \sin^3 \phi \, d\phi$$

$\int_0^{2\pi} \cos^2 \theta \, d\theta$	$= \pi$	(Previously Found)

$$\int_0^{\pi} \sin^3 \phi \, d\phi \;=\; \int_0^{\pi} (1 - \cos^2 \phi) \sin \phi \, d\phi$$

$$= \int_0^{\pi} (\sin \phi - \cos^2 \phi \sin \phi) \, d\phi$$

$$= \int_0^{\pi} \sin \phi \, d\phi \;-\; \int_0^{\pi} \cos^2 \phi \sin \phi \, d\phi$$

$$= [- \cos \phi \,]_0^{\pi} \;+\; \int u^2 \, du \quad , \quad u = \cos \phi$$

$$= -[\cos \phi \,]_0^{\pi} \;+\; \left[\frac{1}{3} u^3 \right]$$

$$= -[\cos \phi \,]_0^{\pi} \;+\; \frac{1}{3}[\cos^3 \phi]_0^{\pi}$$

$$= -[(-1) - (1)] \;+\; \frac{1}{3}[(-1)^3 - (1)^3]$$

$$= [2] + \frac{1}{3}[-2] \;=\; 2 - \frac{2}{3} \;=\; \frac{4}{3}$$

Surface Integrals
For Parametric Surfaces -- Ex. 1e

Compute the surface integral: $\iint_S x^2 \ dS$

Where S is a unit sphere: $\qquad x^2 + y^2 + z^2 = 1$

$\iint_S f(x,y,z) \quad = \quad \int_0^{2\pi} \cos^2 \theta \ d\theta \ \int_0^{\pi} \sin^3 \phi \ d\phi$

Previously Found	$\int_0^{2\pi} \cos^2 \theta \ d\theta \quad = \quad \pi$
	$\int_0^{\pi} \sin^3 \phi \ d\phi \quad = \quad \frac{4}{3}$

$\iint_S f(x,y,z) \ = \ (\pi)\left(\frac{4}{3}\right) \ = \ \frac{4\pi}{3} \ \approx \ 4.19$

Desmos Check:

$$\int_0^{2\pi} \cos^2 \theta \ d\theta \ \cdot \ \int_0^{\pi} \left(1 - \cos^2 \phi\right) \sin \phi \ d\phi$$

$$= \ 4.18879020$$

Surface Integrals

For Parametric Surfaces -- Ex. 2a

Compute the surface integral: $\iint_S y \; dS$

Where S is the surface: $z = x + y^2$

With: $0 \le x \le 1$ and $0 \le y \le 2$

$\iint_S f(x,y,z) \; dS$

$= \iint_D f(x,y,g(x,y)) \sqrt{\left(\frac{\delta z}{\delta x}\right)^2 + \left(\frac{\delta z}{\delta y}\right)^2 + 1} \;\; dA$

$\iint_S y \; dS = \iint_D y \sqrt{1 + \left(\frac{\delta z}{\delta x}\right)^2 + \left(\frac{\delta z}{\delta y}\right)^2} \; dA$

$= \int_0^1 \int_0^2 y \sqrt{1 + (1)^2 + (2y)^2} \; dy \, dx$

$= \int_0^1 \int_0^2 y \sqrt{2 + 4y^2} \; dy \, dx$

$= \sqrt{2} \int_0^1 \int_0^2 y \sqrt{1 + 2y^2} \; dy \, dx$

$= \sqrt{2} \int_0^1 dx \int_0^2 y \sqrt{1 + 2y^2} \; dy$

(Stewart, Calculus Early Transcendentals, p. 1125)

Surface Integrals
For Parametric Surfaces -- Ex. 2b

Compute the surface integral: $\iint_S y \; dS$

Where S is the surface: $z = x + y^2$

With: $0 \le x \le 1$ and $0 \le y \le 2$

$\iint_S y \; dS \;\; = \;\; \sqrt{2} \int_0^1 dx \; \int_0^2 y \sqrt{1 + 2y^2} \; dy$

$\sqrt{2} \int_0^1 dx \;=\; \sqrt{2} \, [\, x \,]_0^1 \;\; = \;\; \sqrt{2}$

$\int_0^2 y \sqrt{1 + 2y^2} \; dy$

$= \frac{1}{4} \int u^{\frac{1}{2}} \; du$ $\qquad \boxed{u = 1 + 2y^2, \quad du = 4y}$

$= \frac{1}{4} \left[\left(\frac{2}{3} \right) u^{\frac{3}{2}} \right] \;=\; \frac{1}{4} \left[\left(\frac{2}{3} \right) (1 + 2y^2)^{\frac{3}{2}} \right]_0^2$

$= \frac{1}{6} \left[(9)^{\frac{3}{2}} - (1)^{\frac{2}{3}} \right] \;=\; \frac{1}{6} [\, 26 \,] \;=\; \frac{13}{3}$

$\iint_S y \; dS \;\; = \;\; \frac{13}{3} \sqrt{2}$

Surface Integrals
For Functions With: $z = g(x, y)$ -- Ex. 3

Given the equation of a torus:

$$\left(\sqrt{x^2 + y^2} - R \right)^2 + z^2 = r^2$$

(a) Find a suitable parameterization of torus.
(b) Compute surface area of torus.

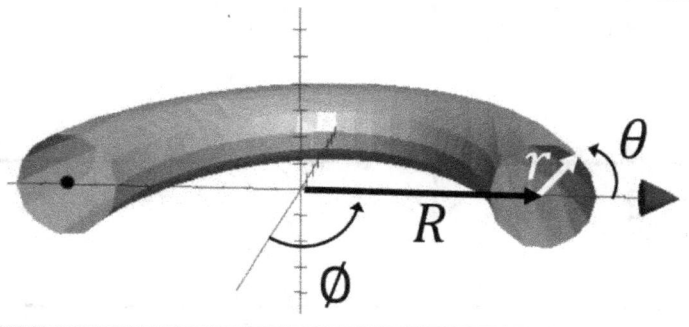

Parameterization	$x = (R + r \cos \theta) \cos \phi$ $y = (R + r \cos \theta) \sin \phi$ $z = r \sin \theta$

Surface Area:

$$A = \int_0^{2\pi} R \int_0^{2\pi} r \; d\theta \; d\phi$$

$$A = Rr \, [\, \theta \,]_0^{2\pi} \, [\, \phi \,]_0^{2\pi} \;\; = \;\; 4\pi^2 \, R \, r$$

Surface Integrals
For Functions With: $z = g(x, y)$ -- Ex. 4a

Show the surface: $z = \dfrac{1}{\sqrt{x^2 + y^2}}$, $1 \le z \le \infty$

Can be filled but not painted.

Parameterization	$x = r \cos \theta$ $y = r \sin \theta$ $z = \dfrac{1}{\sqrt{r^2}} = \dfrac{1}{r}$
$0 \le r \le 1$	$z = \infty \quad \rightarrow \quad r = 0$ $z = 1 \quad \rightarrow \quad r = 1$

$V = \int_0^{2\pi} \int_0^1 \int_1^{\frac{1}{r}} r \; dz \, dr \, d\theta$

$V = \int_0^{2\pi} \int_0^1 r \, [\, z \,]_1^{\frac{1}{r}} \; dr \, d\theta$

$V = \int_0^{2\pi} \int_0^1 r \left[\dfrac{1}{r} - 1 \right] \; dr \, d\theta$

$V = \int_0^{2\pi} \int_0^1 (1 - r) \; dr \, d\theta$

$V = (2\pi) \left[r - \dfrac{r^2}{2} \right]_0^1 = (2\pi) \left[\dfrac{1}{2} \right] = \pi$

Volume is finite so it can be filled.

Surface Integrals
For Functions With: $z = g(x, y)$ -- Ex. 4b

Show the surface: $z = \dfrac{1}{\sqrt{x^2 + y^2}}$, $1 \le z \le \infty$

Can be filled but not painted.

$z_x = -\dfrac{1}{2}(x^2 + y^2)^{\frac{3}{2}}(2x) = -\dfrac{r\cos\theta}{r^{\frac{3}{2}}}$

$z_x = -\dfrac{\cos\theta}{\sqrt{r}}$

$z_y = -\dfrac{1}{2}(x^2 + y^2)^{\frac{3}{2}}(2y) = -\dfrac{r\sin\theta}{r^{\frac{3}{2}}}$

$z_y = -\dfrac{\sin\theta}{\sqrt{r}}$

$A = \iint_D \sqrt{1 + (z_x)^2 + \left(z_y\right)^2}\ dA$

$A = \int_0^{2\pi} \int_0^1 \sqrt{1 + \dfrac{\cos^2\theta}{r} + \dfrac{\sin^2\theta}{r}}\ r\ dr\ d\theta$

$A = \int_0^{2\pi} \int_0^1 \sqrt{1 + \dfrac{1}{r}}\ r\ dr\ d\theta$

$\dfrac{1}{0} = \infty$ is Divergent

$\rightarrow$ Infinite Area

Area is infinite so it cannot be painted.

Surface Integrals
For Functions With: $z = g(x, y)$ -- Ex. 5a

Evaluate: $\iint_S (x^2 + y^2)\, z \;\; dS$

S is the part of the plane: $z = 4 + x + y$

That lies inside the cylinder: $x^2 + y^2 = 4$

$x = r\cos\theta$	$f(x,y,z) = (x^2 + y^2)\, z$
$y = r\sin\theta$	$f(r,\theta) = r^2\, z$
$r = x^2 + y^2$	

$S(x,y,z) = \langle x, y, z \rangle$

$S(x,y) = \langle x, y, 4 + x + y \rangle$

$S(r,\theta) = \langle r\cos\theta, r\sin\theta, 4 + r\cos\theta + r\sin\theta \rangle$

$S_\theta = \langle -r\sin\theta, r\cos\theta, -r\sin\theta + r\cos\theta \rangle$

$S_r = \langle \cos\theta, \sin\theta, \cos\theta + \sin\theta \rangle$

$S_\theta \times S_r$

$$= \begin{vmatrix} i & j & k \\ -r\sin\theta & r\cos\theta & -r\sin\theta + r\cos\theta \\ \cos\theta & \sin\theta & \cos\theta + \sin\theta \end{vmatrix}$$

Surface Integrals

For Functions With: $z = g(x, y)$ -- **Ex. 5b**

Evaluate: $\iint_S (x^2 + y^2) z \ \ dS$

S is the part of the plane: $z = 4 + x + y$

That lies inside the cylinder: $x^2 + y^2 = 4$

$S_\theta \times S_r$

$$= \begin{vmatrix} i & j & k \\ -r \sin \theta & r \cos \theta & -r \sin \theta + r \cos \theta \\ \cos \theta & \sin \theta & \cos \theta + \sin \theta \end{vmatrix}$$

$$= \langle r, r, -r \rangle$$

$\| S_\theta \times S_r \| \ = \ \sqrt{r^2 + r^2 + r^2} \ = \ r\sqrt{3}$

$\iint_S f(x, y, z) \ dS$

$= \iint_D f(f(r, \theta)) \ \| S_\theta \times S_r \| \ dA$

$= \iint_D (r^2 \ z) \ (r\sqrt{3}) \ dA$

$= \int_0^{2\pi} \int_0^2 (r^3 \sqrt{3} \ z) \ dA$

$= \int_0^{2\pi} \int_0^2 (r^3)(4 + r \cos \theta + r \sin \theta) \ dA$

$= \sqrt{3} \int_0^{2\pi} \int_0^2 (4r^3 + r^4 \cos \theta + r^4 \sin \theta) \ dA$

Surface Integrals
For Functions With: $z = g(x, y)$ -- Ex. 5c

Evaluate: $\iint_S (x^2 + y^2) z \ dS$

S is the part of the plane: $z = 4 + x + y$

That lies inside the cylinder: $x^2 + y^2 = 4$

$\iint_S f(x, y, z) \ dS$

$= \sqrt{3} \int_0^{2\pi} \int_0^2 (4r^3 + r^4 \cos\theta + r^4 \sin\theta) \ dr d\theta$

$= \sqrt{3} \int_0^{2\pi} \left[r^4 + \dfrac{r^5}{5} \cos\theta + \dfrac{r^5}{5} \sin\theta \right]_0^2 \ d\theta$

$= \sqrt{3} \int_0^{2\pi} \left[16 + \dfrac{32}{5} \cos\theta + \dfrac{32}{5} \sin\theta \right] \ d\theta$

$= 16\sqrt{3} \int_0^{2\pi} \left[1 + \dfrac{2}{5} \cos\theta + \dfrac{2}{5} \sin\theta \right] \ d\theta$

$= 16\sqrt{3} \left[\theta + \dfrac{2}{5} \sin\theta - \dfrac{2}{5} \cos\theta \right]_0^{2\pi}$

$= 16\sqrt{3} \left[\left(2\pi - \dfrac{2}{5} \right) - \left(-\dfrac{2}{5} \right) \right]$

$= 16\sqrt{3} \left[2\pi \right] \quad = \quad 32\pi\sqrt{3}$

Surface Integrals

For Functions With: $z = g(x, y)$ -- Ex. 6a

Evaluate: $\iint_S (y) \, dS$

S is the surface: $z = 2x + y^2$

Over the square: $[0, 1] \times [0, 1]$

Solution: Use this equation.

$$\iint_S f(x, y, z) \, dS = \iint_D f(r(r, v)) \cdot |r_u \times r_v| \, dA$$
$$= \iint_D f(x, y, z) \cdot |r_x \times r_y| \, dA$$

$r(x, y, z) = \langle x, \, y, \, z \rangle$

$r(x, y) = \langle x, \, y, \, 2x + y^2 \rangle$

$r_x = \langle 1, 0, 2 \rangle \qquad\qquad r_y = \langle 0, 1, 2y \rangle$

$$r_x \times r_y = \begin{vmatrix} i & j & k \\ 1 & 0 & 2 \\ 0 & 1 & 2y \end{vmatrix} = \langle 2, -2y, 1 \rangle$$

$|r_x \times r_y| = \sqrt{4 + 4y^2 + 1} = \sqrt{5 + 4y^2}$

$\iint_S (y) \, dS = \iint_D y \sqrt{5 + 4y^2} \, dA$

Surface Integrals
For Functions With: $z = g(x, y)$ -- Ex. 6b

Evaluate: $\iint_S (y)\, dS$

S is the surface: $z = 2x + y^2$

Over the square: $[0,1] \times [0,1]$

$\iint_S (y)\, dS = \iint_D y\sqrt{5 + 4y^2}\, dA$

$= \int_0^1 \int_0^1 y\sqrt{5 + 4y^2}\, dy\, dx$

$= \frac{1}{8} \int_0^1 \int u^{\frac{1}{2}}\, du\, dx$

$\boxed{\begin{array}{l} u = 5 + 4y^2 \\ du = 8y\, dy \end{array}}$

$= \frac{1}{8} \int_0^1 \left[\frac{2}{3} u^{\frac{3}{2}} \right] dx$

$= \frac{1}{12} \int_0^1 \left[(5 + 4y^2)^{\frac{3}{2}} \right]_0^1 dx$

$= \frac{1}{12} \int_0^1 \left[(5 + 4)^{\frac{3}{2}} - (5 + 0)^{\frac{3}{2}} \right] dx$

$= \frac{1}{12} \int_0^1 \left[27 - 5\sqrt{5} \right] dx$

$= \frac{1}{12} \left[27 - 5\sqrt{5} \right](1) = \frac{1}{12} \left[27 - 5\sqrt{5} \right]$

Stokes' Theorem

Stokes Theorem

$$\int_C F \cdot dr \;=\; \iint_S \text{curl } F \cdot dS$$

Where:

S is a smooth surface bounded by

C a smooth curve w. positive orientation

F is a vector field w. partial derivs. in $\mathbb{R}^3$

Stokes Theorem relates a surface integral over a surface S to a line integral around the boundary Curve of S.

Recall: For $F = \langle P, Q, R \rangle$

$$\text{curl } F \;=\; \nabla \times F$$

$$\text{curl } F \;=\; \begin{vmatrix} i & j & k \\ \dfrac{\delta}{\delta x} & \dfrac{\delta}{\delta y} & \dfrac{\delta}{\delta z} \\ P & Q & R \end{vmatrix}$$

Stokes Theorem -- Ex. 1a

Evaluate $\int_C F \cdot dr$

Where: $F(x, y, z) = \langle -y^2, x, z^2 \rangle$

C is the intersection of the plane: $y + z = 2$

And cylinder: $x^2 + y^2 = 1$

Curve C is an ellipse that makes a circle projection (Domain) on the xy plane

$x^2 + y^2 \leq 1$

$r \leq 1$

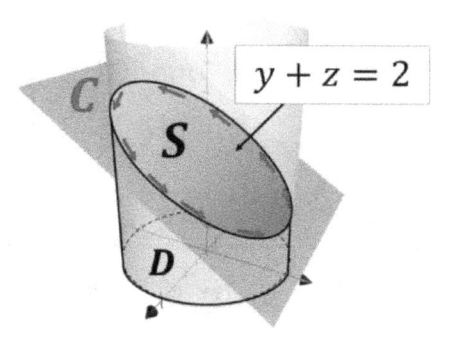

$$curl\ F = \begin{vmatrix} i & j & k \\ \frac{\delta}{\delta x} & \frac{\delta}{\delta y} & \frac{\delta}{\delta z} \\ -y^2 & x & z^2 \end{vmatrix} = (1 + 2y)\ k$$

The projection D of S onto the xy plane is the disk:

$x^2 + y^2 \leq 1 \qquad$ and $\qquad z = 2 - y$

(Stewart, Calculus Early Transcendentals, p. 1136)

Stokes Theorem -- Ex. 1b

Evaluate $\int_C F \cdot dr$

Where: $F(x, y, z) = \langle -y^2, x, z^2 \rangle$

C is the intersection of the plane: $y + z = 2$

And cylinder: $x^2 + y^2 = 1$

Previously Found:

$curl\ F = (1 + 2y)\ \mathbf{k}$

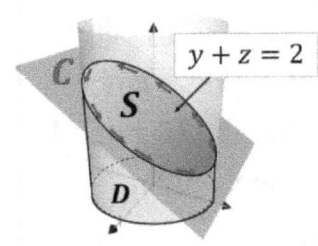

$\int_C F \cdot dr = \iint_S curl\ F \cdot dS$

$= \iint_D (1 + 2y)\ dA$

$= \int_0^{2\pi} \int_0^1 (1 + 2\,r \sin \theta)\ r\ dr\ d\theta$

$= \int_0^{2\pi} \left[\frac{r^2}{2} + \frac{2r^2}{3} \sin \theta \right]_0^1 d\theta$

$= \int_0^{2\pi} \left(\frac{1}{2} + \frac{2}{3} \sin \theta \right)\ d\theta = \frac{1}{2}(2\pi) + 0 = \pi$

Stokes Theorem -- Ex. 2a

Evaluate $\int_C F \cdot dr$ for the given surface.

Where: $F(x, y, z) = \langle x, y, z \rangle$

S is the unit upper hemisphere.

$$x^2 + y^2 + z^2 = 1, \quad z \geq 0$$

$$\delta S = \text{circle:} \quad x^2 + y^2 = 1$$

Evaluate this surface integral by 2 methods:

(a) Directly   and   (b) Using Stokes' Theorem

$r = 1$ $r^2 = 1$	$x^2 + y^2 + z^2 = 1$ $z = \sqrt{1 - (x^2 + y^2)}$ $z = \sqrt{1 - r^2} = \sqrt{1 - 1} = 0$
Find $r(t)$	$r(x, y, z) = \langle x, y, 0 \rangle$ $r(t) = \langle r\cos t, r\sin t, 0 \rangle$ $r'(t) = \langle -\sin t, \cos t, 0 \rangle$
Find $F(r(t))$	$F(x, y, z) = \langle x, y, z \rangle$ $F(r(t)) = \langle \cos t, \sin t, 0 \rangle$

Stokes Theorem -- Ex. 2b

Evaluate $\int_C F \cdot dr$ for the given surface.

Where: $F(x, y, z) = \langle x, y, z \rangle$

           S is the unit upper hemisphere.

           $x^2 + y^2 + z^2 = 1, \;\; z \geq 0$

           $\delta S = $ circle: $x^2 + y^2 = 1$

Evaluate this surface integral by 2 methods:

(a) Directly   and   (b) Using Stokes' Theorem

Previously found	$r(t) = \langle \cos t, \sin t, 0 \rangle$
	$r'(t) = \langle -\sin t, \cos t, 0 \rangle$

a) Direct evaluation:

$\iint_C F \cdot dr = \int_0^{2\pi} F(r(t)) \cdot r'(t) \; dt$

$\iint_C F \cdot dr$

$= \int_0^{2\pi} \langle \cos t, \sin t, 0 \rangle \cdot \langle -\sin t, \cos t, 0 \rangle \, dt$

$= \int_0^{2\pi} (-\cos t \sin t + \cos t \sin t) \; dt$

$= [-\sin t + \sin t]_0^{2\pi} = 0$

Stokes Theorem -- Ex. 2c

Evaluate $\int_C F \cdot dr$ for the given surface.

Where: $F(x, y, z) = \langle x, y, z \rangle$

 S is the unit upper hemisphere.

 $x^2 + y^2 + z^2 = 1, \quad z \geq 0$

 $\delta S = $ circle: $x^2 + y^2 = 1$

Evaluate this surface integral by 2 methods:

(a) Directly   and   (b) Using Stokes' Theorem

(b) Using Stokes' Theorem:

$$\iint_C F \cdot dr = \iint_S curl\ F \cdot dS$$

$F(x, y, z) = \langle x, y, z \rangle$

$$curl\ F = \begin{vmatrix} i & j & k \\ \frac{\delta}{\delta x} & \frac{\delta}{\delta y} & \frac{\delta}{\delta z} \\ x & y & z \end{vmatrix} = \langle 0, 0, 0 \rangle$$

$$\iint_C F \cdot dr = \iint_S curl\ F \cdot dS$$

$$= \iint_S 0\ dS = 0 \qquad \text{Same Answer}$$

The Divergence Theorem

The Divergence Theorem

$$\iint_S F \cdot dS \;=\; \iiint_E div\, F \; dV$$

Where:

E is a simple solid region

S is the boundary surface of E

F is a vector field w. partial derivs. in E

The Divergence Theorem states that the flux of F across the boundary surface of E is equal to the triple integral of the divergence of F over E .

Recall: For $F = \langle P, Q, R \rangle$

$$div\, F \;=\; \frac{\partial P}{\partial x} + \frac{\partial Q}{\partial y} + \frac{\partial R}{\partial z}$$

$$div\, F \;=\; \nabla \cdot F \qquad\qquad \text{(Scalar)}$$

The Divergence Theorem -- Ex. 1

Find the flux of the vector field:

$$F(x, y, z) = \langle z, \ y, \ x \rangle$$

Over the unit sphere: $x^2 + y^2 + z^2 = 1$

First, compute the divergence of F

Recall, for $F = \langle P, Q, R \rangle$

$$div \ F \ = \ \frac{\partial P}{\partial x} + \frac{\partial Q}{\partial y} + \frac{\partial R}{\partial z}$$

Here: $div \ F \ = \ 0 + 1 + 0 \ = \ 1$

The unit sphere is the boundary of the unit ball B

given by: $x^2 + y^2 + z^2 \ \leq \ 1$

$$
\begin{aligned}
\iint_S F \cdot dS \ &= \ \iiint_E div \ F \ \ dV \\
&= \ \iiint_B \ (1) \ \ dV \\
&= \ V(B) \ = \ \frac{4}{3} \pi r^3 \ = \ \frac{4}{3} \pi \ (1)^3 \\
&= \ \frac{4}{3} \ \pi
\end{aligned}
$$

(Stewart, Calculus Early Transcendentals, p. 1143)

The Divergence Theorem -- Ex. 2a

Evaluate $\iint_S F \cdot dS$

Where: $F(x, y, z) = \langle xy, y^2 + e^{xz^2}, \sin xy \rangle$

And S is the surface of region E bounded

by the parabolic cylinder: $z = 1 - x^2$

and the planes: $z = 0$, $y = 0$, and $y + z = 2$

Region E

$-1 \le x \le 1$

$0 \le z \le 1$

$0 \le y \le 2 - z$

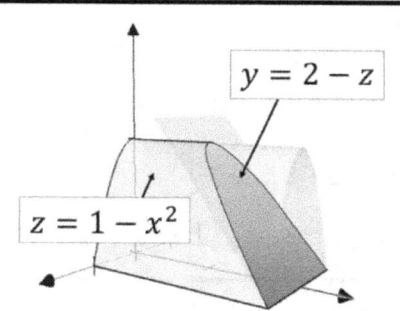

$y = 2 - z$

$z = 1 - x^2$

First, compute the divergence of F

Recall, for $F = \langle P, Q, R \rangle$

$$div\ F = \frac{\partial P}{\partial x} + \frac{\partial Q}{\partial y} + \frac{\partial R}{\partial z}$$

Here: $div\ F = y + 2y + 0 = 3y$

$\iint_S F \cdot dS = \iiint_E div\ F\ dV = \iiint_E 3y\ dV$

(Stewart, Calculus Early Transcendentals, p. 1143)

The Divergence Theorem -- Ex. 2b

Evaluate $\iint_S F \cdot dS$

Where: $F(x, y, z) = \langle\, xy,\ y^2 + e^{xz^2},\ \sin xy\,\rangle$

And S is the surface of region E bounded

by the parabolic cylinder: $z = 1 - x^2$

and the planes: $z = 0$, $y = 0$, and $y + z = 2$

Region E

$-1 \le x \le 1$

$0 \le z \le 1$

$0 \le y \le 2 - z$

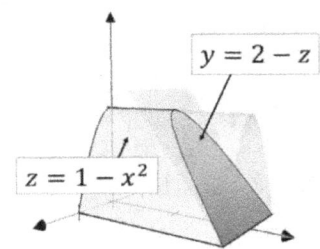

$$\iint_S F \cdot dS = \iiint_E div\ F\ dV = \iiint_E 3y\ dV$$

$$= 3 \int_{-1}^{1} \int_0^{1-x^2} \int_0^{2-z} y\ dy\, dz\, dx$$

$$= 3 \int_{-1}^{1} \int_0^{1-x^2} \left[\frac{y^2}{2}\right]_0^{2-z} dz\, dx$$

$$= \frac{3}{2} \int_{-1}^{1} \int_0^{1-x^2} (2-z)^2\ dz\, dx$$

$$= \frac{3}{2} \int_{-1}^{1} \left[-\frac{(2-z)^3}{3}\right]_0^{1-x^2} dx$$

The Divergence Theorem -- Ex. 2c

Evaluate $\iint_S F \cdot dS$

Where: $F(x, y, z) = \langle xy, y^2 + e^{xz^2}, \sin xy \rangle$

And S is the surface of region E bounded

by the parabolic cylinder: $z = 1 - x^2$

and the planes: $z = 0$, $y = 0$, and $y + z = 2$

$\iint_S F \cdot dS = \iiint_E div\, F \; dV$ (Continued...)

$= -\frac{1}{2} \int_{-1}^{1} [\, (2 - z)^3 \,]_0^{1-x^2} \; dx$

$= -\frac{1}{2} \int_{-1}^{1} [\, (1 + x^2)^3 - (2)^3 \,] \; dx$

$= -\frac{1}{2} \int_{-1}^{1} [\, (1 + 2x^2 + x^4)(1 + x^2) - 8 \,] \; dx$

$= -\frac{1}{2} \int_{-1}^{1} (\, x^6 + 3x^4 + 3x^2 - 7 \,) \; dx$

$= - \int_{0}^{1} (\, x^6 + 3x^4 + 3x^2 - 7 \,) \; dx$

$= - \left[\frac{x^7}{7} + \frac{3x^5}{5} + \frac{3x^3}{3} - 7x \right]_0^1$

$= - \left[\frac{1}{7} + \frac{3}{5} + \frac{3}{3} - 7 \right] = \frac{184}{35}$

The Divergence Theorem -- Ex. 3

Evaluate $\iint_S F \cdot dS$,

Where $F(x, y, z) = \langle x, y, z \rangle$

And S is the surface consisting of the graph of

$z = \sqrt{1 - x^2 - y^2}$ with $z \geq 0$

Give the surface an outer normal.

Solution: Use the Divergence Theorem.

$\iint_S F \cdot dS = \iiint_E (div\ F)\ dV$

$F = \langle x, y, z \rangle$

$div\ F = \frac{\partial}{\partial x}(x) + \frac{\partial}{\partial y}(y) + \frac{\partial}{\partial z}(z)$

$div\ F = 1 + 1 + 1 = 3$

$x^2 + y^2 + z^2 = 1 \quad \rightarrow \quad$ Surface is a sphere.

$\iint_S F \cdot dS = \iiint_E (div\ F)\ dV$

$= \iiint_E (3)\ dV = 3 \iiint_E 1\ dV$

$= 3\ (\text{Vol. of a sphere}) = 3\left(\frac{4}{3}\pi r^3\right) = 4\pi$

The Divergence Theorem -- Ex. 4

Evaluate $\iint_S F \cdot dS$,

Where $F(x, y, z) = \langle 2x, -2y, z^2 \rangle$

And S is the surface of the cylinder

of $x^2 + y^2 = 4$ with $z \in [0, 1]$

Solution: Use the Divergence Theorem.

$\iint_S F \cdot dS = \iiint_E (div\ F)\ dV$

$F = \langle 2x, -2y, z^2 \rangle$

$div\ F = \frac{\partial}{\partial x}(2x) + \frac{\partial}{\partial y}(-2y) + \frac{\partial}{\partial z}(z^2)$

$div\ F = 2 - 2 + 2z = 2z$

$\iint_S F \cdot dS = \iiint_E (div\ F)\ dV$

$= \iiint_E (2z)\ dV = 2 \iiint_E z\ dV$

$= 2 \int_0^{2\pi} \int_0^2 \int_0^1 (z)\ dz\ dr\ d\theta$

$= 2 \int_0^{2\pi} \int_0^2 \left[\frac{1}{2} z^2 \right]_0^1\ dr\ d\theta$

$= \int_0^{2\pi} \int_0^2 [1]\ dr\ d\theta = (2\pi)(2) = 4\pi$

The Divergence Theorem -- Ex. 5

Evaluate $\iint_S F \cdot dS$,

Where $F(x,y,z) = \langle 4xz, -y^2, yz \rangle$

And S is the surface of the cube

Bounded by: $x \in [0,1]$, $y \in [0,1]$, $z \in [0,1]$

Solution: Use the Divergence Theorem.

$\iint_S F \cdot dS = \iiint_E (div\, F)\, dV$

$F = \langle 4xz, -y^2, yz \rangle$

$div\, F = \frac{\partial}{\partial x}(4xz) + \frac{\partial}{\partial y}(-y^2) + \frac{\partial}{\partial z}(yz)$

$div\, F = 4z - 2y + y = 4z - y$

$\iint_S F \cdot dS = \iiint_E (div\, F)\, dV$

$= \int_0^1 \int_0^1 \int_0^1 (4z - y)\, dz\, dy\, dx$

$= \int_0^1 \int_0^1 [2z^2 - yz]_0^1\, dy\, dx$

$= \int_0^1 \int_0^1 [2 - y]\, dy\, dx$

$= \int_0^1 \left[2y - \frac{1}{2}y^2\right]_0^1 dx = \int_0^1 \left[\frac{3}{2}\right] dx = \frac{3}{2}$

<u>Vector Calculus Summary</u>

Summary of Vector Calculus	
Fund. Theorem of Calculus $\int_a^b F'(x)\,dx = F(b) - f(a)$	$a \qquad b$
Fund. Thrm. of Line Integrals $\int_C \nabla f \cdot dr$ $\quad = f(r(b)) - f(r(a))$	$r(b)$ $r(a)$ C
Green's Theorem $\iint_D \left(\frac{\delta Q}{\delta x} - \frac{\delta P}{\delta y} \right) dA$ $\quad = \int_C P\,dx + Q\,dy$	C D
Stokes' Theorem $\iint_S curl\, F \cdot dS = \int_C F \cdot dr$	n C S
Divergence Theorem $\iiint_E div\, F\, dV = \iint_S F \cdot dS$	n S E n

(Stewart, Calculus Early Transcendentals, p. 1147)

Summary of Vector Calculus Some Important Equations	
For $\boldsymbol{F} = \langle P, Q, R \rangle$	
$div\ \boldsymbol{F}$	$= \boldsymbol{\nabla} \cdot \boldsymbol{F}$ (Scalar) $= \langle \frac{\partial}{\partial x}, \frac{\partial}{\partial y}, \frac{\partial}{\partial z} \rangle \cdot \langle P, Q, R \rangle$ $= \frac{\partial P}{\partial x} + \frac{\partial Q}{\partial y} + \frac{\partial R}{\partial z}$
$curl\ \boldsymbol{F}$	$= \boldsymbol{\nabla} \times \boldsymbol{F}$ (Vector) $= \begin{vmatrix} i & j & k \\ \frac{\delta}{\delta x} & \frac{\delta}{\delta y} & \frac{\delta}{\delta z} \\ P & Q & R \end{vmatrix}$
$\nabla f(x, y, z)$	$=$ Gradient of f $= \langle f_x, f_y, f_z \rangle$ (Vector)

<u>Tables (Trig, Derivatives, Integrals)</u>

TABLE -- Trig Formulas

$$\sin(u \pm v) = \sin u \cos v \pm \cos u \sin v$$

$$\cos(u \pm v) = \cos u \cos v \mp \sin u \sin v$$

$$\tan(u \pm v) = \frac{\tan u \pm \tan v}{1 \mp \tan u \cdot \tan v}$$

$\sin 2u = 2 \sin u \cos u$	$\tan 2u = \dfrac{2 \tan u}{1 - \tan^2 u}$

$$\cos 2u = \cos^2 u - \sin^2 u = 2\cos^2 u - 1 = 1 - 2\sin^2 u$$

$\sin^2 u = \dfrac{1 - \cos 2u}{2}$	$\cos^2 u = \dfrac{1 + \cos 2u}{2}$	$\tan^2 u = \dfrac{1 - \cos 2u}{1 + \cos 2u}$

$$\sin u + \sin v = 2 \sin\left(\frac{u+v}{2}\right)\cos\left(\frac{u-v}{2}\right)$$

$$\sin u - \sin v = 2 \cos\left(\frac{u+v}{2}\right)\sin\left(\frac{u-v}{2}\right)$$

$$\cos u + \cos v = 2 \cos\left(\frac{u+v}{2}\right)\cos\left(\frac{u-v}{2}\right)$$

$$\cos u - \cos v = -2 \sin\left(\frac{u+v}{2}\right)\sin\left(\frac{u-v}{2}\right)$$

$$\sin u \cdot \sin v = \frac{1}{2}\left[\cos(u-v) - \cos(u+v)\right]$$

$$\cos u \cdot \cos v = \frac{1}{2}\left[\cos(u-v) + \cos(u+v)\right]$$

$$\sin u \cdot \cos v = \frac{1}{2}\left[\sin(u+v) + \sin(u-v)\right]$$

$$\cos u \cdot \sin v = \frac{1}{2}\left[\sin(u+v) - \sin(u-v)\right]$$

TABLE -- Derivatives

$\dfrac{d}{dx}[cu] = cu'$	$\dfrac{d}{dx}[u \pm v] = u' \pm v'$
$\dfrac{d}{dx}[uv] = u'v + uv'$	$\dfrac{d}{dx}\left[\dfrac{u}{v}\right] = \dfrac{vu' - v'u}{v^2}$
$\dfrac{d}{dx}[c] = 0$	$\dfrac{d}{dx}[u^n] = n\,u^{n-1}u'$
$\dfrac{d}{dx}[x] = 1$	$\dfrac{d}{dx}[\,|u|\,] = \dfrac{u}{|u|}\,u'$
$\dfrac{d}{dx}[\ln u] = \dfrac{u'}{u}$	$\dfrac{d}{dx}[e^u] = e^u\,u'$
$\dfrac{d}{dx}[\log_a u] = \dfrac{u'}{(\ln a)u}$	$\dfrac{d}{dx}[a^u] = (\ln a)\,a^u\,u'$
$\dfrac{d}{dx}[\sin u] = (\cos u)\,u'$	$\dfrac{d}{dx}[\cos u] = -(\sin u)\,u'$
$\dfrac{d}{dx}[\tan u] = (\sec^2 u)\,u'$	$\dfrac{d}{dx}[\cot u] = -(\csc^2 u)\,u'$
$\dfrac{d}{dx}[\sec u] = (\sec u \cdot \tan u)\,u'$	$\dfrac{d}{dx}[\csc u] = -(\csc u \cot u)\,u'$
$\dfrac{d}{dx}[\sin^{-1} u] = \dfrac{u'}{\sqrt{1 - u^2}}$	$\dfrac{d}{dx}[\cos^{-1} u] = \dfrac{-u'}{\sqrt{1-u^2}}$
$\dfrac{d}{dx}[\tan^{-1} u] = \dfrac{u'}{1 + u^2}$	$\dfrac{d}{dx}[\cot^{-1} u] = \dfrac{-u'}{1 + u^2}$
$\dfrac{d}{dx}[\sec^{-1} u] = \dfrac{u'}{|u|\sqrt{1 - u^2}}$	$\dfrac{d}{dx}[\csc^{-1} u] = \dfrac{-u'}{|u|\sqrt{1 - u^2}}$

TABLE -- **Integrals** (Constant of Integration not included)	
$\int [f(u) \pm g(u)] \, du = \int f(u) \, du \pm \int g(u) \, du$	
$\int k \cdot f(u) \, du = k \int f(u) \, du$	$\int 1 \, du = u$
$\int u^n \, du = \dfrac{u^{n+1}}{n+1}$	$\int \dfrac{1}{u} \, du = \ln|u|$
$\int e^u \, du = e^u$	$\int a^u \, du = \left(\dfrac{1}{\ln a}\right) a^u$
$\int \sin u \, du = -\cos u$	$\int \cos u \, du = \sin u$
$\int \tan u \, du = -\ln|\cos u|$	$\int \cot u \, du = \ln|\sin u|$
$\int \sec u \, du = \ln|\sec u + \cot u|$	$\int \sec^2 u \, du = \tan u$
$\int \csc u \, du = -\ln|\sec u + \cot u|$	$\int \csc^2 u \, du = -\cot u$
$\int \sec u \cdot \tan u \, du = \sec u$	$\int \csc u \cdot \cot u \, du = -\csc u$
$\int \dfrac{du}{\sqrt{a^2 - u^2}} = \sin^{-1}\left(\dfrac{u}{a}\right)$	$\int \dfrac{du}{a^2 + u^2} = \left(\dfrac{1}{a}\right)\tan^{-1}\left(\dfrac{u}{a}\right)$
$\int \dfrac{du}{u\sqrt{u^2 - a^2}} = \left(\dfrac{1}{a}\right)\sec^{-1}\left(\dfrac{|u|}{a}\right)$	$\int \ln x \, dx = x \ln|x| - x$
$\int \sec u \, du \;=\; \ln|\sec u + \cot u|$	
$\int \sec^2 u \, du \;=\; \tan u$	
$\int \sec^3 u \, du \;=\; \dfrac{1}{2}\left[\sec u \cdot \tan u + \ln|\sec u + \tan u|\right]$	

References

References

- Calculus Early Transcendentals,
  Eighth Edition, 2015, James Stewart.

- Calculus 10e, Tenth Edition, 2014,
  Ron Larson, Bruce Edwards.

- Calculus Early Transcendentals Single Variable,
  Ninth Edition, 2009,
  Howard Anton, Irl Bivens, Stephen Davis.

- Differential Equations With Applications: Class
  Notes With Detailed Examples, 2019,
  Jigarkumar Patel and Kathryn Paulk.

Other Books by Kathryn Paulk

Other Books by Kathryn Paulk

- Algebra 1 Help
- Algebra 2 Help
- Pre-Calculus and Trig Help
- College Algebra Help
- Fractions for Everyone

- Calculus 1 Review in Bite-Size Pieces
- Calculus 2 Review in Bite-Size Pieces
- Calculus 3 Review in Bite-Size Pieces
- Differential Equations With Applications: Class Notes With Examples

- One-Page Summaries for Algebra, Geometry & Pre-Calc.
- Pre-Calculus and Trig Problems & Solutions
- Graphing Functions Using Transformations for Algebra & Pre-Calculus
- Complex Numbers and Polar Curves For Pre-Calc and Trig: With Problems and Detailed Solutions
- Discrete and Continuous Probability Distributions: A Creative Comparison (V2)

- Teach Your Child to SWIM

BIG MATH For Little Kids

Workbooks for Young Children
& Solution Manuals for Parents

- Introduction to Numbers
  (Ages 2 – 5)

- Introduction to Fractions by Sharing Things
  (Ages 3 – 8)

- Introduction to Counting & Fractions by Cooking Breakfast
  (Ages 5 and up)

- Learn About Fractions by Baking Cookies
  (Ages 8 and up)

- Adding Big Numbers, Guessing Numbers and Secret Codes
  (Ages 8 and up)

- Learn to Graph by Riding Bikes on Graph Paper
  (Ages 10 and up)

These books are based on the math activities
Kathy did with her son when he was young.

www.ingramcontent.com/pod-product-compliance
Lightning Source LLC
Chambersburg PA
CBHW070846290526
45795CB00001B/1